Student Solutions Manual f
DIFFERENTIAL EQUATIONS
with Boundary-Value Problems
Fourth Edition

WARREN S. WRIGHT
Loyola Marymount University
and
CAROL D. WRIGHT

PWS Publishing Company

Brooks/Cole Publishing Company

I(T)P® An International Thomson Publishing Company

Pacific Grove • Albany • Bonn • Boston • Cincinnati • Detroit • London • Madrid • Melbourne
Mexico City • New York • Paris • San Francisco • Singapore • Tokyo • Toronto • Washington

Sponsoring Editor: *Gary Ostedt*
Assistant Editor: *Beth Wilbur*
Marketing Team: *Patrick Farrant and Deborah Petit*
Editorial Assistant: *Nancy Conti*
Production Coordinator: *Mary Vezilich*
Cover Design: *Jennifer Mackres*
Printing and Binding: *Malloy Lithographing*

For more information, contact:

BROOKS/COLE PUBLISHING COMPANY
511 Forest Lodge Road
Pacific Grove, CA 93950
USA

International Thomson Editores
Campos Eliseos 385, Piso 7
Col. Polanco
11560 México D. F. México

International Thomson Publishing Europe
Berkshire House 168-173
High Holborn
London WC1V 7AA
England

International Thomson Publishing GmbH
Königswinterer Strasse 418
53227 Bonn
Germany

Thomas Nelson Australia
102 Dodds Street
South Melbourne, 3205
Victoria, Australia

International Thomson Publishing Asia
221 Henderson Road
#05-10 Henderson Building
Singapore 0315

Nelson Canada
1120 Birchmount Road
Scarborough, Ontario
Canada M1K 5G4

International Thomson Publishing Japan
Hirakawacho Kyowa Building, 3F
2-2-1 Hirakawacho
Chiyoda-ku, Tokyo 102
Japan

Printed in the United States of America

10 9 8 7 6 5 4 3

ISBN 0-534-95579-7

Table of Contents

1 Introduction to Differential Equations

Exercises 1.1

3. First-order; nonlinear because of yy'.

6. Second-order; nonlinear because of $\sin y$.

9. Third-order; linear.

12. From $y = 8$ we obtain $y' = 0$, so that $y' + 4y = 0 + 4(8) = 32$.

15. From $y = 5\tan 5x$ we obtain $y' = 25\sec^2 5x$. Then

$$y' = 25\sec^2 5x = 25\left(1 + \tan^2 5x\right) = 25 + (5\tan 5x)^2 = 25 + y^2.$$

18. First write the differential equation in the form $2xy + \left(x^2 + 2y\right)y' = 0$. Implicitly differentiating $x^2y + y^2 = c_1$ we obtain $2xy + \left(x^2 + 2y\right)y' = 0$.

21. Implicitly differentiating $y^2 = c_1\left(x + \frac{1}{4}c_1\right)$ we obtain $y' = c_1/2y$. Then

$$2xy' + y(y')^2 = \frac{c_1x}{y} + \frac{c_1^2}{4y} = \frac{y^2}{y} = y.$$

24. Differentiating $P = ac_1e^{at}/\left(1 + bc_1e^{at}\right)$ we obtain

$$\frac{dP}{dt} = \frac{\left(1 + bc_1e^{at}\right)a^2c_1e^{at} - ac_1e^{at}\cdot abc_1e^{at}}{\left(1 + bc_1e^{at}\right)^2}$$

$$= \frac{ac_1e^{at}}{1 + bc_1e^{at}}\cdot\frac{\left[a\left(1 + bc_1e^{at}\right) - abc_1e^{at}\right]}{1 + bc_1e^{at}} = P(a - bP).$$

27. First write the differential equation in the form $y' = \dfrac{-x^2 - y^2}{x^2 - xy}$. Then $c_1(x + y)^2 = xe^{y/x}$ implies $c_1 = \dfrac{xe^{y/x}}{(x + y)^2}$ and implicit differentiation gives $2c_1(x + y)(1 + y') = xe^{y/x}\dfrac{xy' - y}{x^2} + e^{y/x}$. Solving for y' we obtain

$$y' = \frac{e^{y/x} - \dfrac{y}{x}e^{y/x} - 2c_1(x + y)}{2c_1(x + y) - e^{y/x}} = \frac{1 - \dfrac{y}{x} - \dfrac{2x}{x + y}}{\dfrac{2x}{x + y} - 1} = \frac{-x^2 - y^2}{x^2 - xy}.$$

30. From $y = e^{2x} + xe^{2x}$ we obtain $\dfrac{dy}{dx} = 3e^{2x} + 2xe^{3x}$ and $\dfrac{d^2y}{dx^2} = 8e^{2x} + 4xe^{2x}$ so that $\dfrac{d^2y}{dx^2} - 4\dfrac{dy}{dx} + 4y = 0$.

33. From $y = \ln|x + c_1| + c_2$ we obtain $y' = \dfrac{1}{x + c_1}$ and $y'' = \dfrac{-1}{(x + c_1)^2}$, so that $y'' + (y')^2 = 0$.

36. From $y = x\cos(\ln x)$ we obtain $y' = -\sin(\ln x) + \cos(\ln x)$ and $y'' = \dfrac{-1}{x}\cos(\ln x) - \dfrac{1}{x}\sin(\ln x)$, so that $x^2y'' - xy' + 2y = 0$.

39. From $y = x^2e^x$ we obtain $y' = x^2e^x + 2xe^x$, $y'' = x^2e^x + 4xe^x 2e^x$, and $y''' = x^2e^x + 6xe^x + 6e^x$, so that $y''' - 3y'' + 3y' - y = 0$.

42. From $y = \begin{cases} 0, & x < 0 \\ x^3, & x \ge 0 \end{cases}$ we obtain $y' = \begin{cases} 0, & x < 0 \\ 3x^2, & x \ge 0 \end{cases}$ so that $(y')^2 = \begin{cases} 0, & x < 0 \\ 9x^4, & x \ge 0. \end{cases}$

45. From $y = e^{mx}$ we obtain $y' = me^{mx}$ and $y'' = m^2e^{mx}$. Then $y'' - 5y' + 6y = 0$ implies

$$m^2e^{mx} - 5me^{mx} + 6e^{mx} = (m-2)(m-3)e^{mx} = 0.$$

Since $e^{mx} > 0$ for all x, $m = 2$ and $m = 3$. Thus $y = e^{2x}$ and $y = e^{3x}$ are solutions.

48. Using $y' = mx^{m-1}$ and $y'' = m(m-1)x^{m-2}$ and substituting into the differential equation we obtain $x^2y'' + 6xy' + 4y = [m(m-1) + 6m + 4]x^m$. The right side will be zero provided m satisfies

$$m(m-1) + 6m + 4 = m^2 + 5m + 4 = (m+4)(m+1) = 0.$$

Thus, $m = -4, -1$ and two solutions of the differential equation on the interval $0 < x < \infty$ are $y = x^{-4}$ and $y = x^{-1}$.

Exercises 1.2

3. For $f(x, y) = \dfrac{y}{x}$ we have $\dfrac{\partial f}{\partial y} = \dfrac{1}{x}$. Thus the differential equation will have a unique solution in any region where $x \neq 0$.

6. For $f(x, y) = \dfrac{x^2}{1 + y^3}$ we have $\dfrac{\partial f}{\partial y} = \dfrac{-3x^2y^2}{(1 + y^3)^2}$. Thus the differential equation will have a unique solution in any region where $y \neq -1$.

9. For $f(x, y) = x^3\cos y$ we have $\dfrac{\partial f}{\partial y} = -x^2\sin y$. Thus the differential equation will have a unique solution in the entire plane.

12. Two solutions are $y = 0$ and $y = x^2$. (Also, any constant multiple of x^2 is a solution.)

15. We identify $f(x, y) = \sqrt{y^2 - 9}$ and $\partial f/\partial y = y^2/\sqrt{y^2 - 9}$. Since $\partial f/\partial y$ is discontinuous for $|y| < 3$, the differential equation is not guaranteed to have a unique solution at $(2, -3)$.

18. (a) Since $1+y^2$ and its partial derivative with respect to y are continuous everywhere in the plane, the differential equation has a unique solution through every point in the plane.

(b) Since $\dfrac{d}{dx}(\tan x) = \sec^2 x = 1 + \tan^2 x$ and $\tan 0 = 0$, $y = \tan x$ satisfies the differential equation and the initial condition. Since $-2 < \pi/2 < 2$ and $\tan x$ is undefined for $x = \pi/2$, $y = \tan x$ is not a solution on the interval $-2 < x < 2$.

(c) Since $\tan x$ is continuous and differentiable on $(-\pi/2, \pi/2)$ and not defined at the endpoints of this interval, this is the largest interval of validity for which $y = \tan x$ is a solution of $y' = 1+y^2$, $y(0) = 0$.

21. Setting $x = 0$ and $y = -1/3$ we have $1/(1 + c_1) = -1/3$ so $c_1 = -4$. A solution of the initial-value problem is $y = 1/(1 - 4e^{-x})$.

24. From the initial conditions we obtain the system

$$c_1 e + c_2 e^{-1} = 0$$

$$c_1 e - c_2 e^{-1} = e.$$

Solving we get $c_1 = \frac{1}{2}$ and $c_2 = -\frac{1}{2}e^2$. A solution of the initial-value problem is $y = \frac{1}{2}e^x - \frac{1}{2}e^{2-x}$.

Exercises 1.3

3. The differential equation is $x'(t) = r - kx(t)$ where $k > 0$.

6. The rate at which salt is entering the tank is

$$R_1 = (3 \text{ gal/min}) \cdot (2 \text{ lb/gal}) = 6 \text{ lb/min}.$$

Since the solution is pumped out at a slower rate, it is accumulating at the rate of $(3-2)$gal/min $=$ 1 gal/min. After t minutes there are $300 + t$ gallons of brine in the tank. The rate at which salt is leaving is

$$R_2 = (2 \text{ gal/min}) \cdot \left(\frac{A}{300+t} \text{ lb/gal}\right) = \frac{2A}{300+t} \text{ lb/min}.$$

The differential equation is

$$\frac{dA}{dt} = 6 - \frac{2A}{300+t}.$$

9. Since $i = \dfrac{dq}{dt}$ and $L\dfrac{d^2q}{dt^2} + R\dfrac{dq}{dt} = E(t)$ we obtain $L\dfrac{di}{dt} + Ri = E(t)$.

12. The differential equation is $\dfrac{dA}{dt} = k_1(M - A) - k_2 A$.

15. The net force acting on the mass is

$$F = ma = m\frac{d^2x}{dt^2} = -k(s+x) + mg = -kx + mg - ks.$$

Since the condition of equilibrium is $mg = ks$, the differential equation is

$$m\frac{d^2x}{dt^2} = -ks.$$

Chapter 1 Review Exercises

3. False; since $y = 0$ is a solution.

6. Third-order; ordinary; nonlinear because of $\sin xy$.

9. From $y = x + \tan x$ we obtain $y' = 1 + \sec^2 x$, and $y'' = 2\sec^2 x \tan x$. Using $1 + \tan^2 x = \sec^2 x$ we have $y' + 2xy = 2 + x^2 + y^2$.

12. From $y = \sin 2x + \cosh 2x$ we obtain $y^{(4)} = 16\sin 2x + 16\cosh 2x$ so that $y^{(4)} - 16y = 0$.

15. $y = \frac{1}{2}x^2$

18. $y = \sqrt{x}$

21. For all values of y, $y^2 - 2y \geq -1$. Avoiding left– and right–hand derivatives, we then must have $x^2 - x - 1 > -1$. That is, $x < 0$ or $x > 1$.

24. From Newton's second law we obtain $m\frac{dv}{dt} = \frac{1}{2}mg - \mu\frac{\sqrt{3}}{2}mg$ or $\frac{dv}{dt} = 16\left(1 - \sqrt{3}\,\mu\right)$.

2 First-Order Differential Equations

Exercises 2.1

In many of the following problems we will encounter an expression of the form $\ln|g(y)| = f(x) + c$. To solve for $g(y)$ we exponentiate both sides of the equation. This yields $|g(y)| = e^{f(x)+c} = e^c e^{f(x)}$ which implies $g(y) = \pm e^c e^{f(x)}$. Letting $c_1 = \pm e^c$ we obtain $g(y) = c_1 e^{f(x)}$.

3. From $dy = -e^{-3x}\,dx$ we obtain $y = \frac{1}{3}e^{-3x} + c$.

6. From $dy = 2xe^{-x}dx$ we obtain $y = -2xe^{-x} - 2e^{-x} + c$.

9. From $\frac{1}{y^3}\,dy = \frac{1}{x^2}\,dx$ we obtain $y^{-2} = \frac{2}{x} + c$.

12. From $\left(\frac{1}{y} + 2y\right)\,dy = \sin x\,dx$ we obtain $\ln|y| + y^2 = -\cos x + c$.

15. From $\frac{y}{2+y^2}\,dy = \frac{x}{4+x^2}\,dx$ we obtain $\ln|2+y^2| = \ln|4+x^2| + c$ or $2 + y^2 = c_1\left(4+x^2\right)$.

18. From $\frac{y^2}{y+1}\,dy = \frac{1}{x^2}\,dx$ we obtain $\frac{1}{2}y^2 - y + \ln|y+1| = -\frac{1}{x} + c$ or $\frac{1}{2}y^2 - y + \ln|y+1| = -\frac{1}{x} + c_1$.

21. From $\frac{1}{S}\,dS = k\,dr$ we obtain $S = ce^{kr}$.

24. From $\frac{1}{N}\,dN = \left(te^{t+2} - 1\right)\,dt$ we obtain $\ln|N| = te^{t+2} - e^{t+2} - t + c$.

27. From $\frac{e^{2y} - y}{e^y}\,dy = -\frac{\sin 2x}{\cos x}\,dx = -\frac{2\sin x\cos x}{\cos x}\,dx$ or $\left(e^y - ye^{-y}\right)dy = -2\sin x\,dx$ we obtain $e^y + ye^{-y} + e^{-y} = 2\cos x + c$.

30. From $\frac{y}{(1+y^2)^{1/2}}\,dy = \frac{x}{(1+x^2)^{1/2}}\,dx$ we obtain $\left(1+y^2\right)^{1/2} = \left(1+x^2\right)^{1/2} + c$.

33. From $\frac{y-2}{y+3}\,dy = \frac{x-1}{x+4}\,dx$ or $\left(1 - \frac{5}{y-3}\right)dy = \left(1 - \frac{5}{x+4}\right)dx$ we obtain

$$y - 5\ln|y-3| = x - 5\ln|x+4| + c \quad \text{or} \quad \left(\frac{x+4}{y-3}\right)^5 = c_1 e^{x-y}.$$

36. From $\sec y\frac{dy}{dx} + \sin x\cos y - \cos x\sin y = \sin x\cos y + \cos x\sin y$ we find $\sec y\,dy = 2\sin y\cos x\,dx$ or $\frac{1}{2\sin y\cos y}\,dy = \csc 2y\,dy = \cos x\,dx$. Then $\frac{1}{2}\ln|\csc 2y - \cot 2y| = \sin x + c$.

39. From $\dfrac{1}{y^2}\,dy = \dfrac{1}{e^x + e^{-x}}\,dx = \dfrac{e^x}{(e^x)^2+1}\,dx$ we obtain $-\dfrac{1}{y} = \tan^{-1} e^x + c$.

42. From $\dfrac{1}{1+(2y)^2}\,dy = \dfrac{-x}{1+(x^2)^2}\,dx$ we obtain

$$\frac{1}{2}\tan^{-1} 2y = -\frac{1}{2}\tan^{-1} x^2 + c \quad \text{or} \quad \tan^{-1} 2y + \tan^{-1} x^2 = c_1.$$

Using $y(1) = 0$ we find $c_1 = \pi/4$. The solution of the initial-value problem is

$$\tan^{-1} 2y + \tan^{-1} x^2 = \frac{\pi}{4}.$$

45. From $\dfrac{1}{x^2+1}\,dx = 4\,dy$ we obtain $\tan^{-1} x = 4y + c$. Using $x(\pi/4) = 1$ we find $c = -3\pi/4$. The solution of the initial-value problem is $\tan^{-1} x = 4y - \dfrac{3\pi}{4}$ or $x = \tan\left(4y - \dfrac{3\pi}{4}\right)$.

48. From $\dfrac{1}{1-2y}\,dy = dx$ we obtain $-\dfrac{1}{2}\ln|1-2y| = x + c$ or $1 - 2y = c_1 e^{-2x}$. Using $y(0) = 5/2$ we find $c_1 = -4$. The solution of the initial-value problem is $1 - 2y = -4e^{-2x}$ or $y = 2e^{-2x} + \dfrac{1}{2}$.

51. By inspection a singular solution is $y = 1$.

54. Separating variables we obtain $\dfrac{dy}{(y-1)^2} = dx$. Then $-\dfrac{1}{y-1} = x + c$ and $y = \dfrac{x+c-1}{x+c}$. Setting $x = 0$ and $y = 1.01$ we obtain $c = -100$. The solution is $y = \dfrac{x-101}{x-100}$.

Exercises 2.2

3. Let $M = 5x + 4y$ and $N = 4x - 8y^3$ so that $M_y = 4 = N_x$. From $f_x = 5x + 4y$ we obtain $f = \dfrac{5}{2}x^2 + 4xy + h(y)$, $h'(y) = -8y^3$, and $h(y) = -2y^4$. The solution is $\dfrac{5}{2}x^2 + 4xy - 2y^4 = c$.

6. Let $M = 4x^3 - 3y\sin 3x - y/x^2$ and $N = 2y - 1/x + \cos 3x$ so that $M_y = -3\sin 3x - 1/x^2$ and $N_x = 1/x^2 - 3\sin 3x$. The equation is not exact.

9. Let $M = y^3 - y^2\sin x - x$ and $N = 3xy^2 + 2y\cos x$ so that $M_y = 3y^2 - 2y\sin x = N_x$. From $f_x = y^3 - y^2\sin x - x$ we obtain $f = xy^3 + y^2\cos x - \dfrac{1}{2}x^2 + h(y)$, $h'(y) = 0$, and $h(y) = 0$. The solution is $xy^3 + y^2\cos x - \dfrac{1}{2}x^2 = c$.

12. Let $M = 2x/y$ and $N = -x^2/y^2$ so that $M_y = -2x/y^2 = N_x$. From $f_x = 2x/y$ we obtain $f = \dfrac{x^2}{y} + h(y)$, $h'(y) = 0$, and $h(y) = 0$. The solution is $x^2 = cy$.

15. Let $M = 1-3/x+y$ and $N = 1-3/y+x$ so that $M_y = 1 = N_x$. From $f_x = 1-3/x+y$ we obtain $f = x - 3\ln|x| + xy + h(y)$, $h'(y) = 1 - \dfrac{3}{y}$, and $h(y) = y - 3\ln|y|$. The solution is $x + y + xy - 3\ln|xy| = c$.

18. Let $M = -2y$ and $N = 5y-2x$ so that $M_y = -2 = N_x$. From $f_x = -2y$ we obtain $f = -2xy+h(y)$, $h'(y) = 5y$, and $h(y) = \frac{5}{2}y^2$. The solution is $-2xy + \frac{5}{2}y^2 = c$.

21. Let $M = 4x^3 + 4xy$ and $N = 2x^2 + 2y - 1$ so that $M_y = 4x = N_x$. From $f_x = 4x^3 + 4xy$ we obtain $f = x^4 + 2x^2y + h(y)$, $h'(y) = 2y - 1$, and $h(y) = y^2 - y$. The solution is $x^4 + 2x^2y + y^2 - y = c$.

24. Let $M = 1/x + 1/x^2 - y/\left(x^2+y^2\right)$ and $N = ye^y + x/\left(x^2+y^2\right)$ so that $M_y = \left(y^2 - x^2\right)/\left(x^2+y^2\right)^2 = N_x$. From $f_x = 1/x + 1/x^2 - y/\left(x^2+y^2\right)$ we obtain $f = \ln|x| - \dfrac{1}{x} - \arctan\left(\dfrac{x}{y}\right) + h(y)$, $h'(y) = ye^y$, and $h(y) = ye^y - e^y$. The solution is

$$\ln|x| - \frac{1}{x} - \arctan\left(\frac{x}{y}\right) + ye^y - e^y = c.$$

27. Let $M = 4y + 2x - 5$ and $N = 6y + 4x - 1$ so that $M_y = 4 = N_x$. From $f_x = 4y + 2x - 5$ we obtain $f = 4xy + x^2 - 5x + h(y)$, $h'(y) = 6y - 1$, and $h(y) = 3y^2 - y$. The general solution is $4xy + x^2 - 5x + 3y^2 - y = c$. If $y(-1) = 2$ then $c = 8$ and the solution of the initial-value problem is $4xy + x^2 - 5x + 3y^2 - y = 8$.

30. Let $M = y^2 + y\sin x$ and $N = 2xy - \cos x - 1/\left(1+y^2\right)$ so that $M_y = 2y + \sin x = N_x$. From $f_x = y^2 + y\sin x$ we obtain $f = xy^2 - y\cos x + h(y)$, $h'(y) = \dfrac{-1}{1+y^2}$, and $h(y) = -\tan^{-1}y$. The general solution is $xy^2 - y\cos x - \tan^{-1}y = c$. If $y(0) = 1$ then $c = -1 - \pi/4$ and the solution of the initial-value problem is $xy^2 - y\cos x - \tan^{-1}y = -1 - \dfrac{\pi}{4}$.

33. Equating $M_y = 4xy + e^x$ and $N_x = 4xy + ke^x$ we obtain $k = 1$.

36. Since $f_x = M(x,y) = y^{1/2}x^{-1/2} + x\left(x^2+y\right)^{-1}$ we obtain $f = 2y^{1/2}x^{1/2} + \frac{1}{2}\ln\left|x^2+y\right| + h(x)$ so that $f_y = y^{-1/2}x^{1/2} + \frac{1}{2}\left(x^2+y\right)^{-1} + h'(x)$. Let $N(x,y) = y^{-1/2}x^{1/2} + \frac{1}{2}\left(x^2+y\right)^{-1}$.

39. Let $M = -x^2y^2\sin x + 2xy^2\cos x$ and $N = 2x^2y\cos x$ so that $M_y = -2x^2y\sin x + 4xy\cos x = N_x$. From $f_y = 2x^2y\cos x$ we obtain $f = x^2y^2\cos x + h(y)$, $h'(y) = 0$, and $h(y) = 0$. The solution of the differential equation is $x^2y^2\cos x = c$.

42. Let $M = \left(x^2 + 2xy - y^2\right)/\left(x^2 + 2xy + y^2\right)$ and $N = \left(y^2 + 2xy - x^2\right)/\left(y^2 + 2xy + x^2\right)$ so that $M_y = -4xy/(x+y)^3 = N_x$. From $f_x = \left(x^2 + 2xy + y^2 - 2y^2\right)/(x+y)^2$ we obtain

$f = x + \dfrac{2y^2}{x+y} + h(y)$, $h'(y) = -1$, and $h(y) = -y$. The solution of the differential equation is $x^2 + y^2 = c(x+y)$.

Exercises 2.3

3. For $y' + 4y = \frac{4}{3}$ an integrating factor is $e^{\int 4dx} = e^{4x}$ so that $\frac{d}{dx}\left[e^{4x}y\right] = \frac{4}{3}e^{4x}$ and $y = \frac{1}{3} + ce^{-4x}$ for $-\infty < x < \infty$.

6. For $y' - y = e^x$ an integrating factor is $e^{-\int dx} = e^{-x}$ so that $\frac{d}{dx}\left[e^{-x}y\right] = 1$ and $y = xe^x + ce^x$ for $-\infty < x < \infty$.

9. For $y' + \frac{1}{x}y = \frac{1}{x^2}$ an integrating factor is $e^{\int(1/x)dx} = x$ so that $\frac{d}{dx}[xy] = \frac{1}{x}$ and $y = \frac{1}{x}\ln x + \frac{c}{x}$ for $0 < x < \infty$.

12. For $\frac{dx}{dy} - x = y$ an integrating factor is $e^{-\int dy} = e^{-y}$ so that $\frac{d}{dy}\left[e^{-y}x\right] = ye^{-y}$ and $x = -y - 1 + ce^y$ for $-\infty < y < \infty$.

15. For $y' + \frac{e^x}{1+e^x}y = 0$ an integrating factor is $e^{\int[e^x/(1+e^x)]dx} = 1 + e^x$ so that $\frac{d}{dx}[1 + e^x y] = 0$ and $y = \frac{c}{1+e^x}$ for $-\infty < x < \infty$.

18. For $y' + (\cot x)y = 2\cos x$ an integrating factor is $e^{\int \cot x\, dx} = \sin x$ so that $\frac{d}{dx}[(\sin x)\, y] = 2\sin x\cos x$ and $y = \sin x + c\csc x$ for $0 < x < \pi$.

21. For $y' + \left(1 + \frac{2}{x}\right)y = \frac{e^x}{x^2}$ an integrating factor is $e^{\int[1+(2/x)]dx} = x^2e^x$ so that $\frac{d}{dx}\left[x^2e^xy\right] = e^{2x}$ and $y = \frac{1}{2}\frac{e^x}{x^2} + \frac{ce^{-x}}{x^2}$ for $0 < x < \infty$.

24. For $y' + \frac{2\sin x}{(1-\cos x)}y = \tan x(1-\cos x)$ an integrating factor is $e^{\int[2\sin x/(1-\cos x)]dx} = (1-\cos x)^2$ so that $\frac{d}{dx}\left[(1-\cos x)^2y\right] = \tan x - \sin x$ and $y(1-\cos x)^2 = \ln|\sec x| + \cos x + c$ for $0 < x < \pi/2$.

27. For $y' + \left(3 + \frac{1}{x}\right)y = \frac{e^{-3x}}{x}$ an integrating factor is $e^{\int[3+(1/x)]dx} = xe^{3x}$ so that $\frac{d}{dx}\left[xe^{3x}y\right] = 1$ and $y = e^{-3x} + \frac{ce^{-3x}}{x}$ for $0 < x < \infty$.

30. For $y' + \frac{2}{x}y = \frac{1}{x}(e^x + \ln x)$ an integrating factor is $e^{\int(2/x)dx} = x^2$ so that $\frac{d}{dx}\left[x^2y\right] = xe^x + x\ln x$ and $x^2y = xe^x - e^x + \frac{x^2}{2}\ln x - \frac{1}{4}x^2 + c$ for $0 < x < \infty$.

33. For $\frac{dx}{dy}+\left(2y+\frac{1}{y}\right)x=2$ an integrating factor is $e^{\int[2y+(1/y)]dy}=ye^{y^2}$ so that $\frac{d}{dy}\left[ye^{y^2}x\right]=2ye^{y^2}$ and $x=\frac{1}{y}+\frac{1}{y}ce^{-y^2}$ for $0<y<\infty$.

36. For $\frac{dP}{dt}+(2t-1)P=4t-2$ an integrating factor is $e^{\int(2t-1)\,dt}=e^{t^2-t}$ so that $\frac{d}{dt}\left[Pe^{t^2-t}\right]=(4t-2)e^{t^2-t}$ and $P=2+ce^{t-t^2}$ for $-\infty<t<\infty$.

39. For $y'+(\cosh x)y=10\cosh x$ an integrating factor is $e^{\int\cosh x\,dx}=e^{\sinh x}$ so that $\frac{d}{dx}\left[e^{\sinh x}y\right]=10(\cosh x)e^{\sinh x}$ and $y=10+ce^{-\sinh x}$ for $-\infty<x<\infty$.

42. For $y'-2y=x\left(e^{3x}-e^{2x}\right)$ an integrating factor is $e^{-\int 2dx}=e^{-2x}$ so that $\frac{d}{dx}\left[e^{-2x}y\right]=xe^{x}-x$ and $y=xe^{3x}-e^{3x}-\frac{1}{2}x^2e^{2x}+ce^{2x}$ for $-\infty<x<\infty$. If $y(0)=2$ then $c=3$ and $y=xe^{3x}-e^{3x}-\frac{1}{2}x^2e^{2x}+3e^{2x}$.

45. For $y'+(\tan x)y=\cos^2 x$ an integrating factor is $e^{\int\tan x\,dx}=\sec x$ so that $\frac{d}{dx}\left[(\sec x)\,y\right]=\cos x$ and $y=\sin x\cos x+c\cos x$ for $-\pi/2<x<\pi/2$. If $y(0)=-1$ then $c=-1$ and $y=\sin x\cos x-\cos x$.

48. For $y'+\left(1+\frac{2}{x}\right)y=\frac{2}{x}e^{-x}$ an integrating factor is $e^{\int(1+2/x)dx}=x^2e^{x}$ so that $\frac{d}{dx}\left[x^2e^{x}y\right]=2x$ and $y=e^{-x}+\frac{c}{x^2}e^{-x}$ for $0<x<\infty$. If $y(1)=0$ then $c=-1$ and $y=e^{-x}-\frac{1}{x^2}e^{-x}$.

51. For $y'+2y=f(x)$ an integrating factor is e^{2x} so that

$$ye^{2x}=\begin{cases}\frac{1}{2}e^{2x}+c_1, & 0\le x\le 3;\\ c_2, & x>3.\end{cases}$$

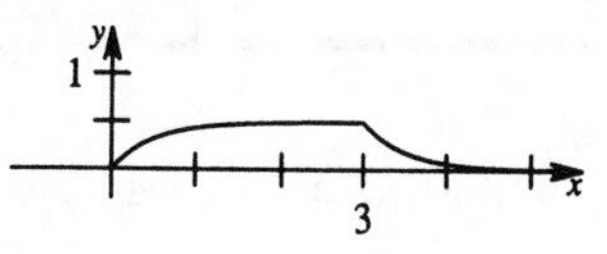

If $y(0)=0$ then $c_1=-1/2$ and for continuity we must have $c_2=\frac{1}{2}e^6-\frac{1}{2}$ so that

$$y=\begin{cases}\frac{1}{2}\left(1-e^{-2x}\right), & 0\le x\le 3;\\ \frac{1}{2}\left(e^6-1\right)e^{-2x}, & x>3.\end{cases}$$

54. For

$$y'+\frac{2x}{1+x^2}y=\begin{cases}\dfrac{x}{1+x^2}, & 0\le x<1;\\ \dfrac{-x}{1+x^2}, & x\ge 1\end{cases}$$

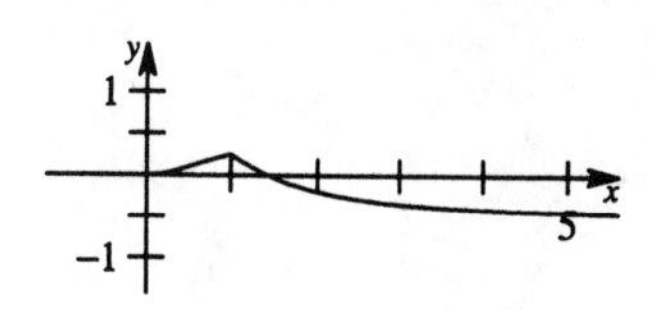

an integrating factor is $1+x^2$ so that

$$\left(1+x^2\right)y = \begin{cases} \frac{1}{2}x^2 + c_1, & 0 \le x < 1; \\ -\frac{1}{2}x^2 + c_2, & x \ge 1. \end{cases}$$

If $y(0) = 0$ then $c_1 = 0$ and for continuity we must have $c_2 = 1$ so that

$$y = \begin{cases} \frac{1}{2} - \frac{1}{2(1+x^2)}, & 0 \le x < 1; \\ \frac{3}{2(1+x^2)} - \frac{1}{2}, & x \ge 1. \end{cases}$$

57. An integrating factor for $y' - 2xy = 1$ is e^{-x^2}. Thus

$$\frac{d}{dx}\left[e^{-x^2}y\right] = e^{-x^2}$$

$$e^{-x^2}y = \int_0^x e^{-t^2}\,dt = \operatorname{erf}(x) + c$$

and

$$y = e^{x^2}\operatorname{erf}(x) + ce^{x^2}.$$

From $y(1) = 1$ we get $1 = e\operatorname{erf}(1) + ce$, so that $c = e^{-1} - \operatorname{erf}(1)$. Thus

$$\begin{aligned} y &= e^{x^2}\operatorname{erf}(x) + (e^{-1} - \operatorname{erf}(1))e^{x^2} \\ &= e^{x^2-1} + e^{x^2}(\operatorname{erf}(x) - \operatorname{erf}(1)). \end{aligned}$$

Exercises 2.4

3. Letting $x = vy$ we have

$$\begin{aligned} vy(v\,dy + y\,dv) + (y - 2vy)\,dy &= 0 \\ vy\,dv + \left(v^2 - 2v + 1\right)dy &= 0 \\ \frac{v\,dv}{(v-1)^2} + \frac{dy}{y} &= 0 \\ \ln|v-1| - \frac{1}{v-1} + \ln|y| &= c \\ \ln\left|\frac{x}{y} - 1\right| - \frac{1}{x/y - 1} + \ln y &= c \\ (x-y)\ln|x-y| - y &= c(x-y). \end{aligned}$$

6. Letting $y = ux$ we have

$$\left(u^2x^2 + ux^2\right) dx + x^2(u\,dx + x\,du) = 0$$

$$\left(u^2 + 2u\right) dx + x\,du = 0$$

$$\frac{dx}{x} + \frac{du}{u(u+2)} = 0$$

$$\ln|x| + \frac{1}{2}\ln|u| - \frac{1}{2}\ln|u+2| = c$$

$$\frac{x^2u}{u+2} = c_1$$

$$x^2\frac{y}{x} = c_1\left(\frac{y}{x} + 2\right)$$

$$x^2y = c_1(y + 2x).$$

9. Letting $y = ux$ we have

$$-ux\,dx + (x + \sqrt{u}\,x)(u\,dx + x\,du) = 0$$

$$(x + x\sqrt{u}\,)\,du + u^{3/2}\,dx = 0$$

$$\left(u^{-3/2} + \frac{1}{u}\right) du + \frac{dx}{x} = 0$$

$$-2u^{-1/2} + \ln|u| + \ln|x| = c$$

$$\ln|y/x| + \ln|x| = 2\sqrt{x/y} + c$$

$$y(\ln|y| - c)^2 = 4x.$$

12. Letting $y = ux$ we have

$$\left(x^2 + 2u^2x^2\right) dx - ux^2(u\,dx + x\,du) = 0$$

$$\left(1 + u^2\right) dx - ux\,du = 0$$

$$\frac{dx}{x} - \frac{u\,du}{1+u^2} = 0$$

$$\ln|x| - \frac{1}{2}\ln\left(1 + u^2\right) = c$$

$$\frac{x^2}{1+u^2} = c_1$$

$$x^4 = c_1\left(y^2 + x^2\right).$$

Using $y(-1)=1$ we find $c_1=1/2$. The solution of the initial-value problem is $2x^4=y^2+x^2$.

15. From $y'+\dfrac{1}{x}y=\dfrac{1}{x}y^{-2}$ and $w=y^3$ we obtain $\dfrac{dw}{dx}+\dfrac{3}{x}w=\dfrac{3}{x}$. An integrating factor is x^3 so that $x^3w=x^3+c$ or $y^3=1+cx^{-3}$.

18. From $y'-\left(1+\dfrac{1}{x}\right)y=y^2$ and $w=y^{-1}$ we obtain $\dfrac{dw}{dx}+\left(1+\dfrac{1}{x}\right)w=-1$. An integrating factor is xe^x so that $xe^xw=-xe^x+e^x+c$ or $y^{-1}=-1+\dfrac{1}{x}+\dfrac{c}{x}e^{-x}$.

21. From $y'-\dfrac{2}{x}y=\dfrac{3}{x^2}y^4$ and $w=y^{-3}$ we obtain $\dfrac{dw}{dx}+\dfrac{6}{x}w=-\dfrac{9}{x^2}$. An integrating factor is x^6 so that $x^6w=-\dfrac{9}{5}x^5+c$ or $y^{-3}=-\dfrac{9}{5}x^{-1}+cx^{-6}$. If $y(1)=\dfrac{1}{2}$ then $c=\dfrac{49}{5}$ and $y^{-3}=-\dfrac{9}{5}x^{-1}+\dfrac{49}{5}x^{-6}$.

24. Let $u=x+y$ so that $du/dx=1+dy/dx$. Then $\dfrac{du}{dx}-1=\dfrac{1-u}{u}$ or $u\,du=dx$. Thus $\dfrac{1}{2}u^2=x+c$ or $u^2=2x+c_1$, and $(x+y)^2=2x+c_1$.

27. Let $u=y-2x+3$ so that $du/dx=dy/dx-2$. Then $\dfrac{du}{dx}+2=2+\sqrt{u}$ or $\dfrac{1}{\sqrt{u}}du=dx$. Thus $2\sqrt{u}=x+c$ and $2\sqrt{y-2x+3}=x+c$.

30. Let $u=3x+2y$ so that $du/dx=3+2\,dy/dx$. Then $\dfrac{du}{dx}=3+\dfrac{2u}{u+2}=\dfrac{5u+6}{u+2}$ and $\dfrac{u+2}{5u+6}du=dx$. Now

$$\frac{u+2}{5u+6}=\frac{1}{5}+\frac{4}{25u+30}$$

so we have

$$\int\left(\frac{1}{5}+\frac{4}{25u+30}\right)du=dx$$

and $\dfrac{1}{5}u+\dfrac{4}{25}\ln|25u+30|=x+c$. Thus

$$\frac{1}{5}(3x+2y)+\frac{4}{25}\ln|75x+50y+30|=x+c.$$

Setting $x=-1$ and $y=-1$ we obtain $c=\frac{4}{5}\ln 95$. The solution is

$$\frac{1}{5}(3x+2y)+\frac{4}{25}\ln|75x+50y+30|=x+\frac{4}{5}\ln 95$$

or

$$5y-5x+2\ln|75x+50y+30|=10\ln 95.$$

Chapter 2 Review Exercises

3. separable, exact, linear in x and y

6. separable, linear in x, Bernoulli

9. Bernoulli

12. exact, linear in y

15. Separating variables we obtain

$$\cos^2 x\,dx = \frac{y}{y^2+1}\,dy \implies \frac{1}{2}x + \frac{1}{4}\sin 2x = \frac{1}{2}\ln\left(y^2+1\right) + c$$

$$\implies 2x + \sin 2x = 2\ln\left(y^2+1\right) + c.$$

18. Write the differential equation in the form $(3y^2+2x)dx+(4y^2+6xy)dy=0$. Letting $M=3y^2+2x$ and $N=4y^2+6xy$ we see that $M_y=6y=N_x$ so the differential equation is exact. From $f_x=3y^2+2x$ we obtain $f=3xy^2+x^2+h(y)$. Then $f_y=6xy+h'(y)=4y^2+6xy$ and $h'(y)=4y^2$ so $h(y)=\frac{4}{3}y^3$. The general solution is

$$3xy^2+x^2+\frac{4}{3}y^3=c.$$

21. Separating variables we obtain

$$y\ln y\,dy = te^t dt \implies \frac{1}{2}y^2\ln|y| - \frac{1}{4}y^2 = te^t - e^t + c.$$

If $y(1)=1$, $c=-1/4$. The solution is $2y^2\ln|y|-y^2=4te^t-4e^t-1$.

24. The differential equation is Bernoulli. Using $w=y^{-1}$ we obtain $-xy^2\dfrac{dw}{dx}+4y=x^4y^2$ or $\dfrac{dw}{dx}-\dfrac{4}{x}w=-x^3$. An integrating factor is x^{-4}, so

$$\frac{d}{dx}\left[x^{-4}w\right] = -\frac{1}{x} \implies x^{-4}w = -\ln x + c$$

$$\implies w = -x^4\ln x + cx^4$$

$$\implies y = \left(cx^4 - x^4\ln x\right)^{-1}.$$

If $y(1)=1$ then $c=1$ and $y=\left(x^4-x^4\ln x\right)^{-1}$.

3 Modeling with First-Order Differential Equations

Exercises 3.1

3. Let $P = P(t)$ be the population at time t. From $dP/dt = kt$ and $P(0) = P_0 = 500$ we obtain $P = 500e^{kt}$. Using $P(10) = 575$ we find $k = \frac{1}{10}\ln 1.15$. Then $P(30) = 500e^{3\ln 1.15} \approx 760$ years.

6. Let $N = N(t)$ be the amount at time t. From $dN/dt = kt$ and $N(0) = 100$ we obtain $N = 100e^{kt}$. Using $N(6) = 97$ we find $k = \frac{1}{6}\ln 0.97$. Then $N(24) = 100e^{(1/6)(\ln 0.97)24} = 100(0.97)^4 \approx 88.5$ mg.

9. Let $I = I(t)$ be the intensity, t the thickness, and $I(0) = I_0$. If $dI/dt = kI$ and $I(3) = 0.25I_0$ then $I = I_0e^{kt}$, $k = \frac{1}{3}\ln 0.25$, and $I(15) = 0.00098I_0$.

12. Assume that $dT/dt = k(T-5)$ so that $T = 5 + ce^{kt}$. If $T(1) = 55°$ and $T(5) = 30°$ then $k = -\frac{1}{4}\ln 2$ and $c = 59.4611$ so that $T(0) = 64.4611°$.

15. Assume $L\,di/dt + Ri = E(t)$, $L = 0.1$, $R = 50$, and $E(t) = 50$ so that $i = \frac{3}{5} + ce^{-500t}$. If $i(0) = 0$ then $c = -3/5$ and $\lim_{t\to\infty} i(t) = 3/5$.

18. Assume $R\,dq/dt + (1/c)q = E(t)$, $R = 1000$, $C = 5\times 10^{-6}$, and $E(t) = 200$. Then $q = \frac{1}{1000} + ce^{-200t}$ and $i = -200ce^{-200t}$. If $i(0) = 0.4$ then $c = -\frac{1}{500}$, $q(0.005) = 0.003$ coulombs, and $i(0.005) = 0.1472$ amps. As $t \to \infty$ we have $q \to \frac{1}{1000}$.

21. From $dA/dt = 4 - A/50$ we obtain $A = 200 + ce^{-t/50}$. If $A(0) = 30$ then $c = -170$ and $A = 200 - 170e^{-t/50}$.

24. From $\dfrac{dA}{dt} = 10 - \dfrac{10A}{500 - (10-5)t} = 10 - \dfrac{2A}{100 - t}$ we obtain $A = 1000 - 10t + c(100 - t)^2$. If $A(0) = 0$ then $c = -\dfrac{1}{10}$. The tank is empty in 100 minutes.

27. **(a)** From $m\,dv/dt = mg - kv$ we obtain $v = gm/k + ce^{-kt/m}$. If $v(0) = v_0$ then $c = v_0 - gm/k$ and the solution of the initial-value problem is

$$v = \frac{gm}{k} + \left(v_0 - \frac{gm}{k}\right)e^{-kt/m}.$$

(b) As $t \to \infty$ the limiting velocity is gm/k.

(c) From $ds/dt = v$ and $s(0) = s_0$ we obtain

$$s = \frac{gm}{k}t - \frac{m}{k}\left(v_0 - \frac{gm}{k}\right)e^{-kt/m} + s_0 + \frac{m}{k}\left(v_0 - \frac{gm}{k}\right).$$

30. Separating variables we obtain

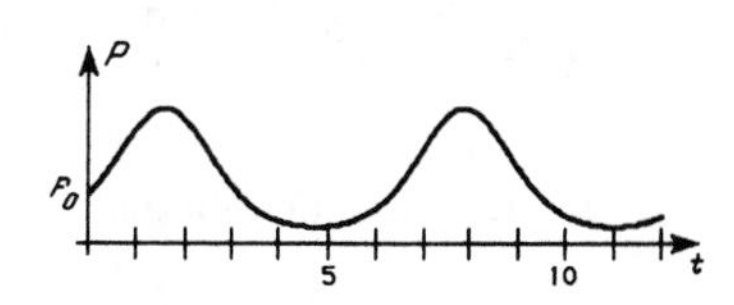

$$\frac{dP}{P} = k\cos t\,dt \implies \ln|P| = k\sin t + c \implies P = c_1 e^{k\sin t}.$$

If $P(0) = P_0$ then $c_1 = P_0$ and $P = P_0 e^{k\sin t}$.

33. **(a)** Letting $t = 0$ correspond to 1790 we have $P(0) = 3.929$ and $P(t) = 3.929e^{kt}$. Using $t = 10$, which corresponds to 1800, we have

$$5.308 = P(10) = 3.929e^{10k}.$$

This implies that $k = 0.03$, so that

$$P(t) = 3.929e^{0.03t}.$$

(b)

Year	Census Population	Predicted Population	Error	% Error
1790	3.929	3.929	0.000	0.00
1800	5.308	5.308	0.000	0.00
1810	7.240	7.171	0.069	0.95
1820	9.638	9.688	-0.050	-0.52
1830	12.866	13.088	-0.222	-1.73
1840	17.069	17.682	-0.613	-3.59
1850	23.192	23.888	-0.696	-3.00
1860	31.433	32.272	-0.839	-2.67
1870	38.558	43.599	-5.041	-13.07
1880	50.156	58.901	-8.745	-17.44
1890	62.948	79.574	-16.626	-26.41
1900	75.996	107.503	-31.507	-41.46
1910	91.972	145.234	-53.262	-57.91
1920	105.711	196.208	-90.497	-85.61
1930	122.775	265.074	-142.299	-115.90
1940	131.669	358.109	-226.440	-171.98
1950	150.697	483.798	-333.101	-221.04

Exercises 3.2

3. From $\dfrac{dP}{dt} = P\left(10^{-1} - 10^{-7}P\right)$ and $P(0) = 5000$ we obtain $P = \dfrac{500}{0.0005 + 0.0995e^{-0.1t}}$ so that $P \to 1{,}000{,}000$ as $t \to \infty$. If $P(t) = 500{,}000$ then $t = 52.9$ months.

6. From Problem 5 we have $P = e^{a/b}e^{-ce^{-bt}}$ so that

$$\frac{dP}{dt} = bce^{a/b-bt}e^{-ce^{-bt}} \quad \text{and} \quad \frac{d^2P}{dt^2} = b^2ce^{a/b-bt}e^{-ce^{-bt}}\left(ce^{-bt} - 1\right).$$

Setting $d^2P/dt^2 = 0$ and using $c = a/b - \ln P_0$ we obtain $t = (1/b)\ln(a/b - \ln P_0)$ and $P = e^{a/b-1}$.

9. If $\alpha \neq \beta$, $\dfrac{dX}{dt} = k(\alpha - X)(\beta - X)$, and $X(0) = 0$ then $\left(\dfrac{1/(\beta-\alpha)}{\alpha - X} + \dfrac{1/(\alpha-\beta)}{\beta - X}\right)dX = k\,dt$ so that

$X = \dfrac{\alpha\beta - \alpha\beta e^{(\alpha-\beta)kt}}{\beta - \alpha e^{(\alpha-\beta)kt}}$. If $\alpha = \beta$ then $\dfrac{1}{(\alpha - X)^2}\,dX = k\,dt$ and $X = \alpha - \dfrac{1}{kt + c}$.

12. In this case the differential equation is

$$\frac{dh}{dt} = -\frac{0.6}{25}\sqrt{h} = -\frac{3}{125}\sqrt{h}.$$

Separating variables and integrating we have $2h^{1/2} = -\dfrac{3}{125}t + c_1$ or $h^{1/2} = -\dfrac{3}{250}t + c_2$. From $h(0) = 20$ we find $c_2 = \sqrt{20}$, so $h = \left(\sqrt{20} - \dfrac{3}{250}t\right)^2$. Solving $h(t) = 0$ for t we find that the tank empties in $\dfrac{250}{3}\sqrt{20}$ s ≈ 6.2 min.

15. (a) Separating variables we obtain

$$\frac{m\,dv}{mg - kv^2} = dt$$

$$\frac{1}{g}\frac{dv}{1 - (kv/mg)^2} = dt$$

$$\frac{\sqrt{mg}}{\sqrt{k}\,g}\frac{\sqrt{k/mg}\,dv}{1 - (\sqrt{k}\,v/\sqrt{mg}\,)^2} = dt$$

$$\sqrt{\frac{m}{kg}}\tanh^{-1}\frac{\sqrt{k}\,v}{\sqrt{mg}} = t + c$$

$$\tanh^{-1}\frac{\sqrt{k}\,v}{\sqrt{mg}} = \sqrt{\frac{kg}{m}}\,t + c_1.$$

Thus the velocity at time t is

$$v(t) = \sqrt{\frac{mg}{k}}\tanh\left(\sqrt{\frac{kg}{m}}\,t + c_1\right).$$

Setting $t = 0$ and $v = v_0$ we find $c_1 = \tanh^{-1}(\sqrt{k}\,v_0/\sqrt{mg}\,)$.

(b) Since $\tanh t \to 1$ as $t \to \infty$, we have $v \to \sqrt{mg/k}$ as $t \to \infty$.

(c) Integrating the expression for $v(t)$ in part (a) we obtain

$$s(t) = \sqrt{\frac{mg}{k}}\int \tanh\left(\sqrt{\frac{kg}{m}}\,t + c_1\right)dt = \frac{m}{k}\ln\left[\cosh\left(\sqrt{\frac{kg}{m}}\,t + c_1\right)\right] + c_2.$$

Setting $t = 0$ and $s = s_0$ we find $c_2 = s_0 - \ln\cosh c_1$.

18. (a) We have $dP/dt = P(a - bP)$ with $P(0) = 3.929$ million. Using separation of variables we obtain

$$P(t) = \frac{3.929a}{3.929b + (a - 3.929b)e^{-at}} = \frac{a/b}{1 + (a/3.929b - 1)e^{-at}}$$

$$= \frac{c}{1 + (c/3.929 - 1)e^{-at}}.$$

At $t = 60(1850)$ the population is 23.192 million, so

$$23.192 = \frac{c}{1 + (c/3.929 - 1)e^{-60a}}$$

or $c = 23.192 + 23.192(c/3.929 - 1)e^{-60a}$. At $t = 120(1910)$

$$91.972 = \frac{c}{1 + (c/3.929 - 1)e^{-120a}}$$

or $c = 91.972 + 91.972(c/3.929 - 1)(e^{-60a})^2$. Combining the two equations for c we get

$$\left(\frac{(c - 23.192)/23.192}{c/3.929 - 1}\right)^2 \left(\frac{c}{3.929} - 1\right) = \frac{c - 91.972}{91.972}$$

or

$$91.972(3.929)(c - 23.192)^2 = (23.192)^2(c - 91.972)(c - 3.929).$$

The solution of this quadratic equation is $c = 197.274$. This in turn gives $a = 0.0313$. Therefore

$$P(t) = \frac{197.274}{1 + 49.21e^{-0.0313t}}.$$

(b)

Year	Census Population	Predicted Population	Error	% Error
1790	3.929	3.929	0.000	0.00
1800	5.308	5.334	-0.026	-0.49
1810	7.240	7.222	0.018	0.24
1820	9.638	9.746	-0.108	-1.12
1830	12.866	13.090	-0.224	-1.74
1840	17.069	17.475	-0.406	-2.38
1850	23.192	23.143	0.049	0.21
1860	31.433	30.341	1.092	3.47
1870	38.558	39.272	-0.714	-1.85
1880	50.156	50.044	0.112	0.22
1890	62.948	62.600	0.348	0.55
1900	75.996	76.666	-0.670	-0.88
1910	91.972	91.739	0.233	0.25
1920	105.711	107.143	-1.432	-1.35
1930	122.775	122.140	0.635	0.52
1940	131.669	136.068	-4.399	-3.34
1950	150.697	148.445	2.252	1.49

21. **(a)** Writing the equation in the form $(x - \sqrt{x^2 + y^2}\,)dx + y\,dy$ we identify $M = x - \sqrt{x^2 + y^2}$ and $N = y$. Since M and N are both homogeneous of degree 1 we use the substitution $y = ux$. It follows that

$$\left(x - \sqrt{x^2 + u^2x^2}\right) dx + ux(u\,dx + x\,du) = 0$$

$$x\left[\left(1 - \sqrt{1 + u^2}\right) + u^2\right] dx + x^2u\,du = 0$$

$$-\frac{u\,du}{1 + u^2 - \sqrt{1 + u^2}} = \frac{dx}{x}$$

$$\frac{u\,du}{\sqrt{1 + u^2}\,(1 - \sqrt{1 + u^2}\,)} = \frac{dx}{x}.$$

Letting $w = 1 - \sqrt{1 + u^2}$ we have $dw = -u\,du/\sqrt{1 + u^2}$ so that

$$-\ln\left(1 - \sqrt{1 + u^2}\right) = \ln x + c$$

$$\frac{1}{1 - \sqrt{1 + u^2}} = c_1x$$

$$1 - \sqrt{1 + u^2} = -\frac{c_2}{x} \qquad (-c_2 = 1/c_1)$$

$$1 + \frac{c_2}{x} = \sqrt{1 + \frac{y^2}{x^2}}$$

$$1 + \frac{2c_2}{x} + \frac{c_2^2}{x^2} = 1 + \frac{y^2}{x^2}.$$

Solving for y^2 we have

$$y^2 = 2c_2x + c_2^2 = 4\left(\frac{c_2}{2}\right)\left(x + \frac{c_2}{2}\right)$$

which is a family of parabolas symmetric with respect to the x-axis with vertex at $(-c_2/2, 0)$ and focus at the origin.

(b) Writing the differential equation as $yy' + x = \sqrt{x^2 + y^2}$ and then squaring and simplifying we obtain $y = 2xy' + y(y')^2$. Let $w = y^2$ and write the differential equation as $y^2 = 2xyy' + y^2(y')^2$. Now $dw/dx = 2yy'$ so $w = xw' + \frac{1}{4}(w')^2$. Using Problem 54 in Exercises 1.1 we obtain $w = cx + \frac{1}{4}c^2$ or $y^2 = cx + (c/2)^2$. Letting $c/2 = c_2$ we have $y^2 = 2c_2x + c_2^2$, which is the solution obtained in part (a).

(c) Let $u = x^2 + y^2$ so that

$$\frac{du}{dx} = 2x + 2y\frac{dy}{dx}.$$

Then
$$y\frac{dy}{dx} = \frac{1}{2}\frac{du}{dx} - x$$
and the differential equation can be written in the form
$$\frac{1}{2}\frac{du}{dx} - x = -x + \sqrt{u} \quad \text{or} \quad \frac{1}{2}\frac{du}{dx} = \sqrt{u}\,.$$
Separating variables and integrating we have
$$\frac{du}{2\sqrt{u}} = dx$$
$$\sqrt{u} = x + c$$
$$u = x^2 + 2cx + c^2$$
$$x^2 + y^2 = x^2 + 2cx + c^2$$
$$y^2 = 2cx + c^2.$$

Exercises 3.3

3. The amounts of x and y are the same at about $t = 5$ days. The amounts of x and z are the same at about $t = 20$ days. The amounts of y and z are the same at about $t = 147$ days. The time when y and z are the same makes sense because most of A and half of B are gone, so half of C should have been formed.

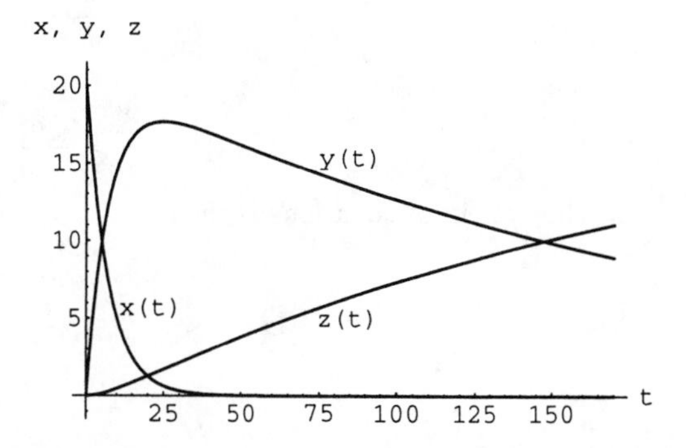

6. Let x_1, x_2, and x_3 be the amounts of salt in tanks A, B, and C, respectively, so that
$$x_1' = \frac{1}{100}x_2 \cdot 2 - \frac{1}{100}x_1 \cdot 6 = \frac{1}{50}x_2 - \frac{3}{50}x_1$$
$$x_2' = \frac{1}{100}x_1 \cdot 6 + \frac{1}{100}x_3 - \frac{1}{100}x_2 \cdot 2 - \frac{1}{100}x_2 \cdot 5 = \frac{3}{50}x_1 - \frac{7}{100}x_2 + \frac{1}{100}x_3$$
$$x_3' = \frac{1}{100}x_2 \cdot 5 - \frac{1}{100}x_3 - \frac{1}{100}x_3 \cdot 4 = \frac{1}{20}x_2 - \frac{1}{20}x_3.$$

Exercises 3.3

9. From the graph we see that the populations are first equal at about $t = 5.6$. The approximate periods of x and y are both 45.

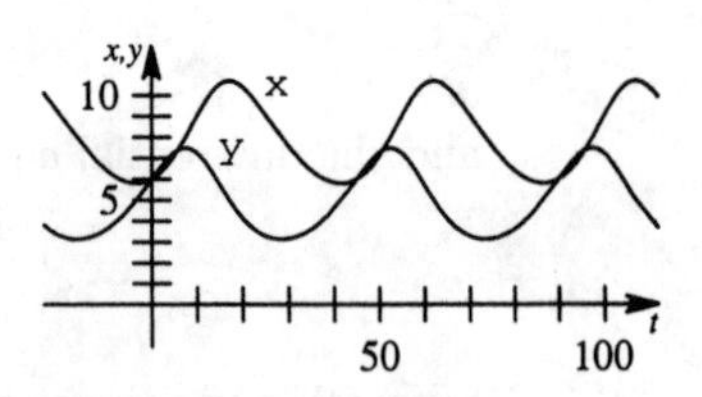

12. By Kirchoff's first law we have $i_1 = i_2 + i_3$. By Kirchoff's second law, on each loop we have $E(t) = Li_1' + R_1 i_2$ and $E(t) = Li_1' + R_2 i_3 + \frac{1}{C}q$ so that $q = CR_1 i_2 - CR_2 i_3$. Then $i_3 = q' = CR_1 i_2' - CR_2 i_3$ so that the system is

$$Li_2' + Li_3' + R_1 i_2 = E(t)$$

$$-R_1 i_2' + R_2 i_3' + \frac{1}{C} i_3 = 0.$$

15. We first note that $s(t) + i(t) + r(t) = n$. Now the rate of change of the number of susceptible persons, $s(t)$, is proportional to the number of contacts between the number of people infected and the number who are susceptible; that is, $ds/dt = -k_1 s_i$. We use $-k_1$ because $s(t)$ is decreasing. Next, the rate of change of the number of persons who have recovered is proportional to the number infected; that is, $dr/dt = k_2 i$ where k_2 is positive since r is increasing. Finally, to obtain di/dt we use

$$\frac{d}{dt}(s + i + r) = \frac{d}{dt} n = 0.$$

This gives

$$\frac{di}{dt} = -\frac{dr}{dt} - \frac{ds}{dt} = -k_2 i + k_1 si.$$

The system of equations is then

$$\frac{ds}{dt} = -k_1 si$$

$$\frac{di}{dt} = -k_2 i + k_1 si$$

$$\frac{dr}{dt} = k_2 i.$$

A reasonable set of initial conditions is $i(0) = i_0$, the number of infected people at time 0, $s(0) = n - i_0$, and $r(0) = 0$.

Chapter 3 Review Exercises

3. From $dE/dt = -E/RC$ and $E(t_1) = E_0$ we obtain $E = E_0 e^{(t_1-t)/RC}$.

6. We first solve $\left(1 - \dfrac{t}{10}\right)\dfrac{di}{dt} + 0.2i = 4$. Separating variables we obtain $\dfrac{di}{40-2i} = \dfrac{dt}{10-t}$. Then

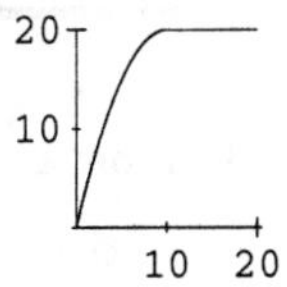

$$-\frac{1}{2}\ln|40-2i| = -\ln|10-t| + c \quad \text{or} \quad \sqrt{40-2i} = c_1(10-t).$$

Since $i(0) = 0$ we must have $c_1 = 2/\sqrt{10}$. Solving for i we get $i(t) = 4t - \frac{1}{5}t^2$, $0 \le t < 10$. For $t \ge 10$ the equation for the current becomes $0.2i = 4$ or $i = 20$. Thus

$$i(t) = \begin{cases} 4t - \frac{1}{5}t^2, & 0 \le t < 10 \\ 20, & t \ge 10 \end{cases}.$$

9. From $\dfrac{dx}{dt} = k_1 x(\alpha - x)$ we obtain $\left(\dfrac{1/\alpha}{x} + \dfrac{1/\alpha}{\alpha - x}\right)dx = k_1\,dt$ so that $x = \dfrac{\alpha c_1 e^{\alpha k_1 t}}{1 + c_1 e^{\alpha k_1 t}}$. From $\dfrac{dy}{dt} = k_2 xy$ we obtain

$$\ln|y| = \frac{k_2}{k_1}\ln\left|1 + c_1 e^{\alpha k_1 t}\right| + c \quad \text{or} \quad y = c_2\left(1 + c_1 e^{\alpha k_1 t}\right)^{k_2/k_1}.$$

4 Differential Equations of Higher-Order

Exercises 4.1

3. From $y = c_1e^{4x} + c_2e^{-x}$ we find $y' = 4c_1e^{4x} - c_2e^{-x}$. Then $y(0) = c_1 + c_2 = 1$, $y'(0) = 4c_1 - c_2 = 2$ so that $c_1 = 3/5$ and $c_2 = 2/5$. The solution is $y = \frac{3}{5}e^{4x} + \frac{2}{5}e^{-x}$.

6. From $y = c_1 + c_2x^2$ we find $y' = 2c_2x$. Then $y(0) = c_1 = 0$, $y'(0) = 2c_2 \cdot 0 = 0$ and $y'(0) = 1$ is not possible. Since $a_2(x) = x$ is 0 at $x = 0$, Theorem 4.1 is not violated.

9. From $y = c_1e^x \cos x + c_2e^x \sin x$ we find $y' = c_1e^x(-\sin x + \cos x) + c_2e^x(\cos x + \sin x)$.

(a) We have $y(0) = c_1 = 1$, $y'(0) = c_1 + c_2 = 0$ so that $c_1 = 1$ and $c_2 = -1$. The solution is $y = e^x \cos x - e^x \sin x$.

(b) We have $y(0) = c_1 = 1$, $y(\pi) = -c_1e^{\pi} = -1$, which is not possible.

(c) We have $y(0) = c_1 = 1$, $y(\pi/2) = c_2e^{\pi/2} = 1$ so that $c_1 = 1$ and $c_2 = e^{-\pi/2}$. The solution is $y = e^x \cos x + e^{-\pi/2}e^x \sin x$.

(d) We have $y(0) = c_1 = 0$, $y(\pi) = -c_1e^{\pi} = 0$ so that $c_1 = 0$ and c_2 is arbitrary. Solutions are $y = c_2e^x \sin x$, for any real numbers c_2.

12. Since $a_0(x) = \tan x$ and $x_0 = 0$ the problem has a unique solution for $-\pi/2 < x < \pi/2$.

15. Since $(-4)x + (3)x^2 + (1)(4x - 3x^2) = 0$ the functions are linearly dependent.

18. Since $(1)\cos 2x + (1)1 + (-2)\cos^2 x = 0$ the functions are linearly dependent.

21. The functions are linearly independent since $W\left(1+x, x, x^2\right) = \begin{vmatrix} 1+x & x & x^2 \\ 1 & 1 & 2x \\ 0 & 0 & 2 \end{vmatrix} = 2 \neq 0.$

24. The functions satisfy the differential equation and are linearly independent since

$$W(\cosh 2x, \sinh 2x) = 2$$

for $-\infty < x < \infty$. The general solution is

$$y = c_1 \cosh 2x + c_2 \sinh 2x.$$

27. The functions satisfy the differential equation and are linearly independent since

$$W\left(x^3, x^4\right) = x^6 \neq 0$$

for $0 < x < \infty$. The general solution is

$$y = c_1x^3 + c_2x^4.$$

30. The functions satisfy the differential equation and are linearly independent since

$$W(1, x, \cos x, \sin x) = 1$$

for $-\infty < x < \infty$. The general solution is

$$y = c_1 + c_2 x + c_3 \cos x + c_4 \sin x.$$

33. The functions $y_1 = e^{2x}$ and $y_2 = e^{5x}$ form a fundamental set of solutions of the homogeneous equation, and $y_p = 6e^x$ is a particular solution of the nonhomogeneous equation.

36. The functions $y_1 = x^{-1/2}$ and $y_2 = x^{-1}$ form a fundamental set of solutions of the homogeneous equation, and $y_p = \frac{1}{15}x^2 - \frac{1}{6}x$ is a particular solution of the nonhomogeneous equation.

Exercises 4.2

In Problems 3-9 we use reduction of order to find a second solution. In Problems 12-24 we use formula (5) from the text.

3. Define $y = u(x)e^{2x}$ so

$$y' = 2ue^{2x} + u'e^{2x}, \quad y'' = e^{2x}u'' + 4e^{2x}u' + 4e^{2x}u, \quad \text{and} \quad y'' - 4y' + 4y = 4e^{2x}u'' = 0.$$

Therefore $u'' = 0$ and $u = c_1 x + c_2$. Taking $c_1 = 1$ and $c_2 = 0$ we see that a second solution is $y_2 = xe^{2x}$.

6. Define $y = u(x)\sin 3x$ so

$$y' = 3u\cos 3x + u'\sin 3x, \quad y'' = u''\sin 3x + 6u'\cos 3x - 9u\sin 3x,$$

and

$$y'' + 9y = (\sin 3x)u'' + 6(\cos 3x)u' = 0 \quad \text{or} \quad u'' + 6(\cot 3x)u' = 0.$$

If $w = u'$ we obtain the first-order equation $w' + 6(\cot 3x)w = 0$ which has the integrating factor $e^{6\int \cot 3x\, dx} = \sin^2 3x$. Now

$$\frac{d}{dx}[(\sin^2 3x)w] = 0 \quad \text{gives} \quad (\sin^2 3x)w = c.$$

Therefore $w = u' = c\csc^2 3x$ and $u = c_1 \cot 3x$. A second solution is $y_2 = \cot 3x \sin 3x = \cos 3x$.

9. Define $y = u(x)e^{2x/3}$ so

$$y' = \frac{2}{3}e^{2x/3}u + e^{2x/3}u', \quad y'' = e^{2x/3}u'' + \frac{4}{3}e^{2x/3}u' + \frac{4}{9}e^{2x/3}u$$

and

$$9y'' - 12y' + 4y = 9e^{2x/3}u'' = 0.$$

Therefore $u'' = 0$ and $u = c_1x + c_2$. Taking $c_1 = 1$ and $c_2 = 0$ we see that a second solution is $y_2 = xe^{2x/3}$.

12. Identifying $P(x) = 2/x$ we have

$$y_2 = x^2 \int \frac{e^{-\int (2/x)\,dx}}{x^4}\,dx = x^2 \int x^{-6}dx = -\frac{1}{5}x^{-3}.$$

A second solution is $y_2 = x^{-3}$.

15. Identifying $P(x) = 2(1+x)/\left(1 - 2x - x^2\right)$ we have

$$y_2 = (x+1) \int \frac{e^{-\int 2(1+x)dx/(1-2x-x^2)}}{(x+1)^2}\,dx = (x+1) \int \frac{e^{\ln(1-2x-x^2)}}{(x+1)^2}\,dx$$

$$= (x+1) \int \frac{1 - 2x - x^2}{(x+1)^2}\,dx = (x+1) \int \left[\frac{2}{(x+1)^2} - 1\right] dx$$

$$= (x+1)\left[-\frac{2}{x+1} - x\right] = -2 - x^2 - x.$$

A second solution is $y_2 = x^2 + x + 2$.

18. Identifying $P(x) = -3/x$ we have

$$y_2 = x^2\cos(\ln x) \int \frac{e^{-\int -3\,dx/x}}{x^4\cos^2(\ln x)}\,dx = x^2\cos(\ln x) \int \frac{x^3}{x^4\cos^2(\ln x)}\,dx$$

$$= x^2\cos(\ln x)\tan(\ln x) = x^2\sin(\ln x).$$

A second solution is $y_2 = x^2\sin(\ln x)$.

21. Identifying $P(x) = -1/x$ we have

$$y_2 = x \int \frac{e^{-\int -dx/x}}{x^2}\,dx = x \int \frac{dx}{x} = x\ln|x|.$$

A second solution is $y_2 = x\ln|x|$.

24. Identifying $P(x) = 1/x$ we have

$$y_2 = \cos(\ln x) \int \frac{e^{-\int dx/x}}{\cos^2(\ln x)}\,dx = \cos(\ln x) \int \frac{1/x}{\cos^2(\ln x)}\,dx = \cos(\ln x)\tan(\ln x) = \sin(\ln x).$$

A second solution is $y_2 = \sin(\ln x)$.

27. Define $y = u(x)e^x$ so

$$y' = ue^x + u'e^x, \quad y'' = u''e^x + 2u'e^x + ue^x$$

and

$$y'' - 3y' + 2y = e^x u'' - e^x u' = 0 \quad \text{or} \quad u'' - u' = 0.$$

If $w = u'$ we obtain the first order equation $w' - w = 0$ which has the integrating factor $e^{-\int dx} = e^{-x}$. Now

$$\frac{d}{dx}[e^{-x}w] = 0 \quad \text{gives} \quad e^{-x}w = c.$$

Therefore $w = u' = ce^x$ and $u = ce^x$. A second solution is $y_2 = e^x e^x = e^{2x}$. To find a particular solution we try $y_p = Ae^{3x}$. Then $y' = 3Ae^{3x}$, $y'' = 9Ae^{3x}$, and $9Ae^{3x} - 3\left(3Ae^{3x}\right) + 2Ae^{3x} = 5e^{3x}$. Thus $A = 5/2$ and $y_p = \frac{5}{2}e^{3x}$. The general solution is

$$y = c_1 e^x + c_2 e^{2x} + \frac{5}{2}e^{3x}.$$

Exercises 4.3

3. From $m^2 - 36 = 0$ we obtain $m = 6$ and $m = -6$ so that $y = c_1 e^{6x} + c_2 e^{-6x}$.

6. From $3m^2 + 1 = 0$ we obtain $m = i/\sqrt{3}$ and $m = -i/\sqrt{3}$ so that $y = c_1 \cos x/\sqrt{3} + c_2 \sin x/\sqrt{3}$.

9. From $m^2 + 8m + 16 = 0$ we obtain $m = -4$ and $m = -4$ so that $y = c_1 e^{-4x} + c_2 x e^{-4x}$.

12. From $m^2 + 4m - 1 = 0$ we obtain $m = -2 \pm \sqrt{5}$ so that $y = c_1 e^{(-2+\sqrt{5})x} + c_2 e^{(-2-\sqrt{5})x}$.

15. From $m^2 - 4m + 5 = 0$ we obtain $m = 2 \pm i$ so that $y = e^{2x}(c_1 \cos x + c_2 \sin x)$.

18. From $2m^2 + 2m + 1 = 0$ we obtain $m = -1/2 \pm i/2$ so that

$$y = e^{-x/2}(c_1 \cos x/2 + c_2 \sin x/2).$$

21. From $m^3 - 1 = 0$ we obtain $m = 1$ and $m = -1/2 \pm \sqrt{3}i/2$ so that

$$y = c_1 e^x + e^{-x/2}\left(c_2 \cos \sqrt{3}x/2 + c_3 \sin \sqrt{3}x/2\right).$$

24. From $m^3 + 3m^2 - 4m - 12 = 0$ we obtain $m = -2$, $m = 2$, and $m = -3$ so that

$$y = c_1 e^{-2x} + c_2 e^{2x} + c_3 e^{-3x}.$$

27. From $m^3 + 3m^2 + 3m + 1 = 0$ we obtain $m = -1$, $m = -1$, and $m = -1$ so that

$$y = c_1 e^{-x} + c_2 x e^{-x} + c_3 x^2 e^{-x}.$$

30. From $m^4 - 2m^2 + 1 = 0$ we obtain $m = 1$, $m = 1$, $m = -1$, and $m = -1$ so that

$$y = c_1 e^x + c_2 x e^x + c_3 e^{-x} + c_4 x e^{-x}.$$

33. From $m^5 - 16m = 0$ we obtain $m = 0$, $m = 2$, $m = -2$, and $m = \pm 2i$ so that

$$y = c_1 + c_2 e^{2x} + c_3 e^{-2x} + c_4 \cos 2x + c_5 \sin 2x.$$

36. From $2m^5 - 7m^4 + 12m^3 + 8m^2 = 0$ we obtain $m = 0$, $m = 0$, $m = -1/2$, and $m = 2 \pm 2i$ so that

$$y = c_1 + c_2 x + c_3 e^{-x/2} + e^{2x}(c_4 \cos 2x + c_5 \sin 2x).$$

39. From $m^2 + 6m + 5 = 0$ we obtain $m = -1$ and $m = -5$ so that $y = c_1 e^{-x} + c_2 e^{-5x}$. If $y(0) = 0$ and $y'(0) = 3$ then $c_1 + c_2 = 0$, $-c_1 - 5c_2 = 3$, so $c_1 = 3/4$, $c_2 = -3/4$, and $y = \frac{3}{4}e^{-x} - \frac{3}{4}e^{-5x}$.

42. From $m^2 - 2m + 1 = 0$ we obtain $m = 1$ and $m = 1$ so that $y = c_1 e^x + c_2 x e^x$. If $y(0) = 5$ and $y'(0) = 10$ then $c_1 = 5$, $c_1 + c_2 = 10$ so $c_1 = 5$, $c_2 = 5$, and $y = 5e^x + 5xe^x$.

45. From $m^2 - 3m + 2 = 0$ we obtain $m = 1$ and $m = 2$ so that $y = c_1 e^x + c_2 e^{2x}$. If $y(1) = 0$ and $y'(1) = 1$ then $c_1 e + c_2 e^2 = 0$, $c_1 e + 2c_2 e^2 = 0$ so $c_1 = -e^{-1}$, $c_2 = e^{-2}$, and $y = -e^{x-1} + e^{2x-2}$.

48. From $m^3 + 2m^2 - 5m - 6 = 0$ we obtain $m = -1$, $m = 2$, and $m = -3$ so that

$$y = c_1 e^{-x} + c_2 e^{2x} + c_3 e^{-3x}.$$

If $y(0) = 0$, $y'(0) = 0$, and $y''(0) = 1$ then

$$c_1 + c_2 + c_3 = 0, \quad -c_1 + 2c_2 - 3c_3 = 0, \quad c_1 + 4c_2 + 9c_3 = 1,$$

so $c_1 = -1/6$, $c_2 = 1/15$, $c_3 = 1/10$, and

$$y = -\frac{1}{6}e^{-x} + \frac{1}{15}e^{2x} + \frac{1}{10}e^{-3x}.$$

51. From $m^4 - 3m^3 + 3m^2 - m = 0$ we obtain $m = 0$, $m = 1$, $m = 1$, and $m = 1$ so that $y = c_1 + c_2 e^x + c_3 x e^x + c_4 x^2 e^x$. If $y(0) = 0$, $y'(0) = 0$, $y''(0) = 1$, and $y'''(0) = 1$ then

$$c_1 + c_2 = 0, \quad c_2 + c_3 = 0, \quad c_2 + 2c_3 + 2c_4 = 1, \quad c_2 + 3c_3 + 6c_4 = 1,$$

so $c_1 = 2$, $c_2 = -2$, $c_3 = 2$, $c_4 = -1/2$, and

$$y = 2 - 2e^x + 2xe^x - \frac{1}{2}x^2 e^x.$$

54. From $m^2 + 4 = 0$ we obtain $m = \pm 2i$ so that $y = c_1 \cos 2x + c_2 \sin 2x$. If $y(0) = 0$ and $y(\pi) = 0$ then $c_1 = 0$ and $y = c_2 \sin 2x$.

57. Using a CAS to solve the auxiliary equation $m^3 - 6m^2 + 2m + 1$ we find $m_1 = -0.270534$, $m_2 = 0.658675$, and $m_3 = 5.61186$. The general solution is

$$y = c_1 e^{-0.270534x} + c_2 e^{0.658675x} + c_3 e^{5.61186x}.$$

60. Using a CAS to solve the auxiliary equation $m^4 + 2m^2 - m + 2 = 0$ we find $m_1 = 1/2 + \sqrt{3}\,i/2$, $m_2 = 1/2 - \sqrt{3}\,i/2$, $m_3 = -1/2 + \sqrt{7}\,i/2$, and $m_4 = -1/2 - \sqrt{7}\,i/2$. The general solution is

$$y = e^{x/2}\left(c_1 \cos\frac{\sqrt{3}}{2}x + c_2 \sin\frac{\sqrt{3}}{2}x\right) + e^{-x/2}\left(c_3 \cos\frac{\sqrt{7}}{2}x + c_4 \sin\frac{\sqrt{7}}{2}x\right).$$

Exercises 4.4

3. From $m^2 - 10m + 25 = 0$ we find $m_1 = m_2 = 5$. Then $y_c = c_1e^{5x} + c_2xe^{5x}$ and we assume $y_p = Ax + B$. Substituting into the differential equation we obtain $25A = 30$ and $-10A + 25B = 3$. Then $A = \frac{6}{5}$, $B = \frac{6}{5}$, $y_p = \frac{6}{5}x + \frac{6}{5}$, and

$$y = c_1e^{5x} + c_2xe^{5x} + \frac{6}{5}x + \frac{6}{5}.$$

6. From $m^2 - 8m + 20 = 0$ we find $m_1 = 2 + 4i$ and $m_2 = 2 - 4i$. Then $y_c = e^{2x}(c_1\cos 4x + c_2\sin 4x)$ and we assume $y_p = Ax^2 + Bx + C + (Dx + E)e^x$. Substituting into the differential equation we obtain

$$2A - 8B + 20C = 0$$
$$-6D + 13E = 0$$
$$-16A + 20B = 0$$
$$13D = -26$$
$$20A = 100.$$

Then $A = 5$, $B = 4$, $C = \frac{11}{10}$, $D = -2$, $E = -\frac{12}{13}$, $y_p = 5x^2 + 4x + \frac{11}{10} + \left(-2x - \frac{12}{13}\right)e^x$ and

$$y = e^{2x}(c_1\cos 4x + c_2\sin 4x) + 5x^2 + 4x + \frac{11}{10} + \left(-2x - \frac{12}{13}\right)e^x.$$

9. From $m^2 - m = 0$ we find $m_1 = 1$ and $m_2 = 0$. Then $y_c = c_1e^x + c_2$ and we assume $y_p = Ax$. Substituting into the differential equation we obtain $-A = -3$. Then $A = 3$, $y_p = 3x$ and $y = c_1e^x + c_2 + 3x$.

12. From $m^2 - 16 = 0$ we find $m_1 = 4$ and $m_2 = -4$. Then $y_c = c_1e^{4x} + c_2e^{-4x}$ and we assume $y_p = Axe^{4x}$. Substituting into the differential equation we obtain $8A = 2$. Then $A = \frac{1}{4}$, $y_p = \frac{1}{4}xe^{4x}$ and

$$y = c_1e^{4x} + c_2e^{-4x} + \frac{1}{4}xe^{4x}.$$

15. From $m^2 + 1 = 0$ we find $m_1 = i$ and $m_2 = -i$. Then $y_c = c_1\cos x + c_2\sin x$ and we assume $y_p = (Ax^2 + Bx)\cos x + (Cx^2 + Dx)\sin x$. Substituting into the differential equation we obtain $4C = 0$, $2A + 2D = 0$, $-4A = 2$, and $-2B + 2C = 0$. Then $A = -\frac{1}{2}$, $B = 0$, $C = 0$, $D = \frac{1}{2}$, $y_p = -\frac{1}{2}x^2\cos x + \frac{1}{2}x\sin x$, and

$$y = c_1\cos x + c_2\sin x - \frac{1}{2}x^2\cos x + \frac{1}{2}x\sin x.$$

Exercises 4.4

18. From $m^2-2m+2=0$ we find $m_1=1+i$ and $m_2=1-i$. Then $y_c=e^x(c_1\cos x+c_2\sin x)$ and we assume $y_p=Ae^{2x}\cos x+Be^{2x}\sin x$. Substituting into the differential equation we obtain $A+2B=1$ and $-2A+B=-3$. Then $A=\frac{7}{5}$, $B=-\frac{1}{5}$, $y_p=\frac{7}{5}e^{2x}\cos x-\frac{1}{5}e^{2x}\sin x$ and

$$y=e^x(c_1\cos x+c_2\sin x)+\frac{7}{5}e^{2x}\cos x-\frac{1}{5}e^{2x}\sin x.$$

21. From $m^3-6m^2=0$ we find $m_1=m_2=0$ and $m_3=6$. Then $y_c=c_1+c_2x+c_3e^{6x}$ and we assume $y_p=Ax^2+B\cos x+C\sin x$. Substituting into the differential equation we obtain $-12A=3$, $6B-C=-1$, and $B+6C=0$. Then $A=-\frac{1}{4}$, $B=-\frac{6}{37}$, $C=\frac{1}{37}$, $y_p=-\frac{1}{4}x^2-\frac{6}{37}\cos x+\frac{1}{37}\sin x$, and

$$y=c_1+c_2x+c_3e^{6x}-\frac{1}{4}x^2-\frac{6}{37}\cos x+\frac{-1}{37}\sin x.$$

24. From $m^3-m^2-4m+4=0$ we find $m_1=1$, $m_2=2$, and $m_3=-2$. Then $y_c=c_1e^x+c_2e^{2x}+c_3e^{-2x}$ and we assume $y_p=A+Bxe^x+Cxe^{2x}$. Substituting into the differential equation we obtain $4A=5$, $-3B=-1$, and $4C=1$. Then $A=\frac{5}{4}$, $B=\frac{1}{3}$, $C=\frac{1}{4}$, $y_p=\frac{5}{4}+\frac{1}{3}xe^x+\frac{1}{4}xe^{2x}$, and

$$y=c_1e^x+c_2e^{2x}+c_3e^{-2x}+\frac{5}{4}+\frac{1}{3}xe^x+\frac{1}{4}xe^{2x}.$$

27. We have $y_c=c_1\cos 2x+c_2\sin 2x$ and we assume $y_p=A$. Substituting into the differential equation we find $A=-\frac{1}{2}$. Thus $y=c_1\cos 2x+c_2\sin 2x-\frac{1}{2}$. From the initial conditions we obtain $c_1=0$ and $c_2=\sqrt{2}$, so $y=\sqrt{2}\sin 2x-\frac{1}{2}$.

30. We have $y_c=c_1e^{-2x}+c_2xe^{-2x}$ and we assume $y_p=(Ax^3+Bx^2)e^{-2x}$. Substituting into the differential equation we find $A=\frac{1}{6}$ and $B=\frac{3}{2}$. Thus $y=c_1e^{-2x}+c_2xe^{-2x}+\left(\frac{1}{6}x^3+\frac{3}{2}x^2\right)e^{-2x}$. From the initial conditions we obtain $c_1=2$ and $c_2=9$, so

$$y=2e^{-2x}+9xe^{-2x}+\left(\frac{1}{6}x^3+\frac{3}{2}x^2\right)e^{-2x}.$$

33. We have $x_c=c_1\cos\omega t+c_2\sin\omega t$ and we assume $x_p=At\cos\omega t+Bt\sin\omega t$. Substituting into the differential equation we find $A=-F_0/2\omega$ and $B=0$. Thus $x=c_1\cos\omega t+c_2\sin\omega t-(F_0/2\omega)t\cos\omega t$. From the initial conditions we obtain $c_1=0$ and $c_2=F_0/2\omega^2$, so

$$x=(F_0/2\omega^2)\sin\omega t-(F_0/2\omega)t\cos\omega t.$$

36. We have $y_c=c_1e^{-2x}+e^x(c_2\cos\sqrt{3}\,x+c_3\sin\sqrt{3}\,x)$ and we assume $y_p=Ax+B+Cxe^{-2x}$. Substituting into the differential equation we find $A=\frac{1}{4}$, $B=-\frac{5}{8}$, and $C=\frac{2}{3}$. Thus

$$y=c_1e^{-2x}+e^x(c_2\cos\sqrt{3}\,x+c_3\sin\sqrt{3}\,x)+\frac{1}{4}x-\frac{5}{8}+\frac{2}{3}xe^{-2x}.$$

From the initial conditions we obtain $c_1=-\frac{23}{12}$, $c_2=-\frac{59}{24}$, and $c_3=\frac{17}{72}\sqrt{3}$, so

$$y=-\frac{23}{12}e^{-2x}+e^x\left(-\frac{59}{24}\cos\sqrt{3}\,x+\frac{17}{72}\sqrt{3}\sin\sqrt{3}\,x\right)+\frac{1}{4}x-\frac{5}{8}+\frac{2}{3}xe^{-2x}.$$

39. We have $y_c = c_1 \cos 2x + c_2 \sin 2x$ and we assume $y_p = A\cos x + B\sin x$ on $[0, \pi/2]$. Substituting into the differential equation we find $A = 0$ and $B = \frac{1}{3}$. Thus $y = c_1 \cos 2x + c_2 \sin 2x + \frac{1}{3}\sin x$ on $[0, \pi/2]$. On $(\pi/2, \infty)$ we have $y = c_3 \cos 2x + c_4 \sin 2x$. From $y(0) = 1$ and $y'(0) = 2$ we obtain

$$c_1 = 1$$

$$\frac{1}{3} + 2c_2 = 2.$$

Solving this system we find $c_1 = 1$ and $c_2 = \frac{5}{6}$. Thus $y = \cos 2x + \frac{5}{6}\sin 2x + \frac{1}{3}\sin x$ on $[0, \pi/2]$. Now continuity of y at $x = \pi/2$ implies

$$\cos\pi + \frac{5}{6}\sin\pi + \frac{1}{3}\sin\frac{\pi}{2} = c_3\cos\pi + c_4\sin\pi$$

or $-1 + \frac{1}{3} = -c_3$. Hence $c_3 = \frac{2}{3}$. Continuity of y' at $x = \pi/2$ implies

$$-2\sin\pi + \frac{5}{3}\cos\pi + \frac{1}{3}\cos\frac{\pi}{2} = -2c_3\sin\pi + 2c_4\cos\pi$$

or $-\frac{5}{3} = -2c_4$. Then $c_4 = \frac{5}{6}$ and the solution of the boundary-value problem is

$$y(x) = \begin{cases} \cos 2x + \frac{5}{6}\sin 2x + \frac{1}{3}\sin x, & 0 \le x \le \pi/2 \\ \frac{2}{3}\cos 2x + \frac{5}{6}\sin 2x, & x > \pi/2 \end{cases}.$$

Exercises 4.5

3. $(D^2 - 4D - 12)y = (D-6)(D+2)y = x - 6$

6. $(D^3 + 4D)y = D(D^2+4)y = e^x \cos 2x$

9. $(D^4 + 8D)y = D(D+2)(D^2 - 2D + 4)y = 4$

12. $(2D-1)y = (2D-1)4e^{x/2} = 8De^{x/2} - 4e^{x/2} = 4e^{x/2} - 4e^{x/2} = 0$

15. D^4 because of x^3

18. $D^2(D-6)^2$ because of x and xe^{6x}

21. $D^3(D^2+16)$ because of x^2 and $\sin 4x$

24. $D(D-1)(D-2)$ because of 1, e^x, and e^{2x}

27. 1, x, x^2, x^3, x^4

30. $D^2 - 9D - 36 = (D-12)(D+3)$; e^{12x}, e^{-3x}

33. $D^3 - 10D^2 + 25D = D(D-5)^2$; 1, e^{5x}, xe^{5x}

36. Applying D to the differential equation we obtain

$$D(2D^2 - 7D + 5)y = 0.$$

Then

$$y = \underbrace{c_1 e^{5x/2} + c_2 e^x}_{y_c} + c_3$$

and $y_p = A$. Substituting y_p into the differential equation yields $5A = -29$ or $A = -29/5$. The general solution is

$$y = c_1 e^{5x/2} + c_2 e^x - \frac{29}{5}.$$

39. Applying D^2 to the differential equation we obtain

$$D^2(D^2 + 4D + 4)y = D^2(D+2)^2 y = 0.$$

Then

$$y = \underbrace{c_1 e^{-2x} + c_2 x e^{-2x}}_{y_c} + c_3 + c_4 x$$

and $y_p = Ax + B$. Substituting y_p into the differential equation yields $4Ax + (4A + 4B) = 2x + 6$. Equating coefficients gives

$$4A = 2$$

$$4A + 4B = 6.$$

Then $A = 1/2$, $B = 1$, and the general solution is

$$y = c_1 e^{-2x} + c_2 x e^{-2x} + \frac{1}{2}x + 1.$$

42. Applying D^4 to the differential equation we obtain

$$D^4(D^2 - 2D + 1)y = D^4(D-1)^2 y = 0.$$

Then

$$y = \underbrace{c_1 e^x + c_2 x e^x}_{y_c} + c_3 x^3 + c_4 x^2 + c_5 x + c_6$$

and $y_p = Ax^3 + Bx^2 + Cx + D$. Substituting y_p into the differential equation yields $Ax^3 + (B - 6A)x^2 + (6A - 4B + C)x + (2B - 2C + D) = x^3 + 4x$. Equating coefficients gives

$$A = 1$$

$$B - 6A = 0$$

$$6A - 4B + C = 4$$

$$2B - 2C + D = 0.$$

Then $A = 1$, $B = 6$, $C = 22$, $D = 32$, and the general solution is

$$y = c_1 e^x + c_2 x e^x + x^3 + 6x^2 + 22x + 32.$$

45. Applying $D(D-1)$ to the differential equation we obtain

$$D(D-1)(D^2-2D-3)y = D(D-1)(D+1)(D-3)y = 0.$$

Then

$$y = \underbrace{c_1e^{3x} + c_2e^{-x}}_{y_c} + c_3e^x + c_4$$

and $y_p = Ae^x + B$. Substituting y_p into the differential equation yields $-4Ae^x - 3B = 4e^x - 9$. Equating coefficients gives $A = -1$ and $B = 3$. The general solution is

$$y = c_1e^{3x} + c_2e^{-x} - e^x + 3.$$

48. Applying $D(D^2+1)$ to the differential equation we obtain

$$D(D^2+1)(D^2+4)y = 0.$$

Then

$$y = \underbrace{c_1\cos 2x + c_2\sin 2x}_{y_c} + c_3\cos x + c_4\sin x + c_5$$

and $y_p = A\cos x + B\sin x + C$. Substituting y_p into the differential equation yields $3A\cos x + 3B\sin x + 4C = 4\cos x + 3\sin x - 8$. Equating coefficients gives $A = 4/3$, $B = 1$, and $C = -2$. The general solution is

$$y = c_1\cos 2x + c_2\sin 2x + \frac{4}{3}\cos x + \sin x - 2.$$

51. Applying $D(D-1)^3$ to the differential equation we obtain

$$D(D-1)^3(D^2-1)y = D(D-1)^4(D+1)y = 0.$$

Then

$$y = \underbrace{c_1e^x + c_2e^{-x}}_{y_c} + c_3x^3e^x + c_4x^2e^x + c_5xe^x + c_6$$

and $y_p = Ax^3e^x + Bx^2e^x + Cxe^x + D$. Substituting y_p into the differential equation yields $6Ax^2e^x + (6A+4B)xe^x + (2B+2C)e^x - D = x^2e^x + 5$. Equating coefficients gives

$$6A = 1$$

$$6A + 4B = 0$$

$$2B + 2C = 0$$

$$-D = 5.$$

Then $A = 1/6$, $B = -1/4$, $C = 1/4$, $D = -5$, and the general solution is

$$y = c_1e^x + c_2e^{-x} + \frac{1}{6}x^3e^x - \frac{1}{4}x^2e^x + \frac{1}{4}xe^x - 5.$$

Exercises 4.5

54. Applying $D^2 - 2D + 10$ to the differential equation we obtain

$$(D^2 - 2D + 10)\left(D^2 + D + \frac{1}{4}\right)y = (D^2 - 2D + 10)\left(D + \frac{1}{2}\right)^2 y = 0.$$

Then

$$y = \underbrace{c_1 e^{-x/2} + c_2 x e^{-x/2}}_{y_c} + c_3 e^x \cos 3x + c_4 e^x \sin 3x$$

and $y_p = Ae^x \cos 3x + Be^x \sin 3x$. Substituting y_p into the differential equation yields $(9B - 27A/4)e^x \cos 3x - (9A + 27B/4)e^x \sin 3x = -e^x \cos 3x + e^x \sin 3x$. Equating coefficients gives

$$-\frac{27}{4}A + 9B = -1$$

$$-9A - \frac{27}{4}B = 1.$$

Then $A = -4/225$, $B = -28/225$, and the general solution is

$$y = c_1 e^{-x/2} + c_2 x e^{-x/2} - \frac{4}{225}e^x \cos 3x - \frac{28}{225}e^x \sin 3x.$$

57. Applying $(D^2 + 1)^2$ to the differential equation we obtain

$$(D^2 + 1)^2(D^2 + D + 1) = 0.$$

Then

$$y = \underbrace{e^{-x/2}\left[c_1 \cos\frac{\sqrt{3}}{2}x + c_2 \sin\frac{\sqrt{3}}{2}x\right]}_{y_c} + c_3 \cos x + c_4 \sin x + c_5 x \cos x + c_6 x \sin x$$

and $y_p = A\cos x + B\sin x + Cx\cos x + Dx\sin x$. Substituting y_p into the differential equation yields

$$(B + C + 2D)\cos x + Dx\cos x + (-A - 2C + D)\sin x - Cx\sin x = x\sin x.$$

Equating coefficients gives

$$B + C + 2D = 0$$

$$D = 0$$

$$-A - 2C + D = 0$$

$$-C = 1.$$

Then $A = 2$, $B = 1$, $C = -1$, and $D = 0$, and the general solution is

$$y = e^{-x/2}\left[c_1 \cos\frac{\sqrt{3}}{2}x + c_2 \sin\frac{\sqrt{3}}{2}x\right] + 2\cos x + \sin x - x\cos x.$$

60. Applying $D(D-1)^2(D+1)$ to the differential equation we obtain

$$D(D-1)^2(D+1)(D^3-D^2+D-1)=D(D-1)^3(D+1)(D^2+1)=0.$$

Then

$$y=\underbrace{c_1e^x+c_2\cos x+c_3\sin x}_{y_c}+c_4+c_5e^{-x}+c_6xe^x+c_7x^2e^x$$

and $y_p=A+Be^{-x}+Cxe^x+Dx^2e^x$. Substituting y_p into the differential equation yields

$$4Dxe^x+(2C+4D)e^x-4Be^{-x}-A=xe^x-e^{-x}+7.$$

Equating coefficients gives

$$\begin{aligned}4D&=1\\ 2C+4D&=0\\ -4B&=-1\\ -A&=7.\end{aligned}$$

Then $A=-7$, $B=1/4$, $C=-1/2$, and $D=1/4$, and the general solution is

$$y=c_1e^x+c_2\cos x+c_3\sin x-7+\frac{1}{4}e^{-x}-\frac{1}{2}xe^x+\frac{1}{4}x^2e^x.$$

63. Applying $D(D-1)$ to the differential equation we obtain

$$D(D-1)(D^4-2D^3+D^2)=D^3(D-1)^3=0.$$

Then

$$y=\underbrace{c_1+c_2x+c_3e^x+c_4xe^x}_{y_c}+c_5x^2+c_6x^2e^x$$

and $y_p=Ax^2+Bx^2e^x$. Substituting y_p into the differential equation yields $2A+2Be^x=1+e^x$. Equating coefficients gives $A=1/2$ and $B=1/2$. The general solution is

$$y=c_1+c_2x+c_3e^x+c_4xe^x+\frac{1}{2}x^2+\frac{1}{2}x^2e^x.$$

66. The complementary function is $y_c=c_1+c_2e^{-x}$. Using D^2 to annihilate x we find $y_p=Ax+Bx^2$. Substituting y_p into the differential equation we obtain $(A+2B)+2Bx=x$. Thus $A=-1$ and $B=1/2$, and

$$\begin{aligned}y&=c_1+c_2e^{-x}-x+\frac{1}{2}x^2\\ y'&=-c_2e^{-x}-1+x.\end{aligned}$$

The initial conditions imply

$$c_1 + c_2 = 1$$
$$-c_2 = 1.$$

Thus $c_1 = 2$ and $c_2 = -1$, and

$$y = 2 - e^{-x} - x + \frac{1}{2}x^2.$$

69. The complementary function is $y_c = c_1 \cos x + c_2 \sin x$. Using $(D^2+1)(D^2+4)$ to annihilate $8\cos 2x - 4\sin x$ we find $y_p = Ax\cos x + Bx\sin x + C\cos 2x + D\sin 2x$. Substituting y_p into the differential equation we obtain $2B\cos x - 3C\cos 2x - 2A\sin x - 3D\sin 2x = 8\cos 2x - 4\sin x$. Thus $A = 2$, $B = 0$, $C = -8/3$, and $D = 0$, and

$$y = c_1\cos x + c_2\sin x + 2x\cos x - \frac{8}{3}\cos 2x$$
$$y' = -c_1\sin x + c_2\cos x + 2\cos x - 2x\sin x + \frac{16}{3}\sin 2x.$$

The initial conditions imply

$$c_2 + \frac{8}{3} = -1$$
$$-c_1 - \pi = 0.$$

Thus $c_1 = -\pi$ and $c_2 = -11/3$, and

$$y = -\pi\cos x - \frac{11}{3}\sin x + 2x\cos x - \frac{8}{3}\cos 2x.$$

72. The complementary function is $y_c = c_1 + c_2x + c_3x^2 + c_4e^x$. Using $D^2(D-1)$ to annihilate $x + e^x$ we find $y_p = Ax^3 + Bx^4 + Cxe^x$. Substituting y_p into the differential equation we obtain $(-6A + 24B) - 24Bx + Ce^x = x + e^x$. Thus $A = -1/6$, $B = -1/24$, and $C = 1$, and

$$y = c_1 + c_2x + c_3x^2 + c_4e^x - \frac{1}{6}x^3 - \frac{1}{24}x^4 + xe^x$$
$$y' = c_2 + 2c_3x + c_4e^x - \frac{1}{2}x^2 - \frac{1}{6}x^3 + e^x + xe^x$$
$$y'' = 2c_3 + c_4e^x - x - \frac{1}{2}x^2 + 2e^x + xe^x.$$
$$y''' = c_4e^x - 1 - x + 3e^x + xe^x$$

The initial conditions imply

$$c_1 + c_4 = 0$$

$$c_2 + c_4 + 1 = 0$$

$$2c_3 + c_4 + 2 = 0$$

$$2 + c_4 = 0.$$

Thus $c_1 = 2$, $c_2 = 1$, $c_3 = 0$, and $c_4 = -2$, and

$$y = 2 + x - 2e^x - \frac{1}{6}x^3 - \frac{1}{24}x^4 + xe^x.$$

Exercises 4.6

The particular solution, $y_p = u_1y_1 + u_2y_2$, in the following problems can take on a variety of forms, especially where trigonometric functions are involved. The validity of a particular form can best be checked by substituting it back into the differential equation.

3. The auxiliary equation is $m^2 + 1 = 0$, so $y_c = c_1 \cos x + c_2 \sin x$ and

$$W = \begin{vmatrix} \cos x & \sin x \\ -\sin x & \cos x \end{vmatrix} = 1.$$

Identifying $f(x) = \sin x$ we obtain

$$u_1' = -\sin^2 x$$

$$u_2' = \cos x \sin x.$$

Then

$$u_1 = \frac{1}{4}\sin 2x - \frac{1}{2}x = \frac{1}{2}\sin x \cos x - \frac{1}{2}x$$

$$u_2 = -\frac{1}{2}\cos^2 x.$$

and

$$y = c_1 \cos x + c_2 \sin x + \frac{1}{2}\sin x \cos^2 x - \frac{1}{2}x \cos x - \frac{1}{2}\cos^2 x \sin x$$

$$= c_1 \cos x + c_2 \sin x - \frac{1}{2}x \cos x$$

for $-\infty < x < \infty$.

6. The auxiliary equation is $m^2+1=0$, so $y_c = c_1\cos x + c_2\sin x$ and

$$W = \begin{vmatrix} \cos x & \sin x \\ -\sin x & \cos x \end{vmatrix} = 1.$$

Identifying $f(x)=\sec^2 x$ we obtain

$$u_1' = -\frac{\sin x}{\cos^2 x}$$

$$u_2' = \sec x.$$

Then

$$u_1 = -\frac{1}{\cos x} = -\sec x$$

$$u_2 = \ln|\sec x + \tan x|$$

and

$$y = c_1\cos x + c_2\sin x - \cos x\sec x + \sin x\ln|\sec x + \tan x|$$

$$= c_1\cos x + c_2\sin x - 1 + \sin x\ln|\sec x + \tan x|$$

for $-\pi/2 < x < \pi/2$.

9. The auxiliary equation is $m^2-4=0$, so $y_c = c_1e^{2x} + c_2e^{-2x}$ and

$$W = \begin{vmatrix} e^{2x} & e^{-2x} \\ 2e^{2x} & -2e^{-2x} \end{vmatrix} = -4.$$

Identifying $f(x)=e^{2x}/x$ we obtain $u_1' = 1/4x$ and $u_2' = -e^{4x}/4x$. Then

$$u_1 = \frac{1}{4}\ln|x|, \qquad u_2 = -\frac{1}{4}\int_{x_0}^{x}\frac{e^{4t}}{t}\,dt$$

and

$$y = c_1e^{2x} + c_2e^{-2x} + \frac{1}{4}\left(e^{2x}\ln|x| - e^{-2x}\int_{x_0}^{x}\frac{e^{4t}}{t}\,dt\right), \quad x_0>0$$

for $x>0$.

12. The auxiliary equation is $m^2-3m+2=(m-1)(m-2)=0$, so $y_c = c_1e^{x} + c_2e^{2x}$ and

$$W = \begin{vmatrix} e^{x} & e^{2x} \\ e^{x} & 2e^{2x} \end{vmatrix} = e^{3x}.$$

Identifying $f(x)=e^{3x}/(1+e^x)$ we obtain

$$u_1' = -\frac{e^{2x}}{1+e^x} = \frac{e^x}{1+e^x} - e^x$$

$$u_2' = \frac{e^x}{1+e^x}.$$

Then $u_1 = \ln(1+e^x) - e^x$, $u_2 = \ln(1+e^x)$, and

$$y = c_1e^x + c_2e^{2x} + e^x\ln(1+e^x) - e^{2x} + e^{2x}\ln(1+e^x)$$

$$= c_1e^x + c_3e^{2x} + (1+e^x)e^x\ln(1+e^x)$$

for $-\infty < x < \infty$.

15. The auxiliary equation is $m^2 - 2m + 1 = (m-1)^2 = 0$, so $y_c = c_1e^x + c_2xe^x$ and

$$W = \begin{vmatrix} e^x & xe^x \\ e^x & xe^x + e^x \end{vmatrix} = e^{2x}.$$

Identifying $f(x) = e^x/\left(1+x^2\right)$ we obtain

$$u_1' = -\frac{xe^xe^x}{e^{2x}\left(1+x^2\right)} = -\frac{x}{1+x^2}$$

$$u_2' = \frac{e^xe^x}{e^{2x}\left(1+x^2\right)} = \frac{1}{1+x^2}.$$

Then $u_1 = -\frac{1}{2}\ln\left(1+x^2\right)$, $u_2 = \tan^{-1}x$, and

$$y = c_1e^x + c_2xe^x - \frac{1}{2}e^x\ln\left(1+x^2\right) + xe^x\tan^{-1}x$$

for $-\infty < x < \infty$.

18. The auxiliary equation is $m^2 + 10m + 25 = (m+5)^2 = 0$, so $y_c = c_1e^{-5x} + c_2xe^{-5x}$ and

$$W = \begin{vmatrix} e^{-5x} & xe^{-5x} \\ -5e^{-5x} & -5xe^{-5x} + e^{-5x} \end{vmatrix} = e^{-10x}.$$

Identifying $f(x) = e^{-10x}/x^2$ we obtain

$$u_1' = -\frac{xe^{-5x}e^{-10x}}{x^2e^{-10x}} = -\frac{e^{-5x}}{x}$$

$$u_2' = \frac{e^{-5x}e^{-10x}}{x^2e^{-10x}} = \frac{e^{-5x}}{x^2}.$$

Then

$$u_1 = -\int_{x_0}^{x}\frac{e^{-5t}}{t}\,dt, \quad x_0 > 0$$

$$u_2 = \int_{x_0}^{x}\frac{e^{-5t}}{t^2}\,dt, \quad x_0 > 0$$

and

$$y = c_1e^{-5x} + c_2xe^{-5x} - e^{-5x}\int_{x_0}^{x}\frac{e^{-5t}}{t}\,dt + xe^{-5x}\int_{x_0}^{x}\frac{e^{-5t}}{t^2}\,dt$$

for $x > 0$.

21. The auxiliary equation is $m^3 + m = m(m^2 + 1) = 0$, so $y_c = c_1 + c_2 \cos x + c_3 \sin x$ and

$$W = \begin{vmatrix} 1 & \cos x & \sin x \\ 0 & -\sin x & \cos x \\ 0 & -\cos x & -\sin x \end{vmatrix} = 1.$$

Identifying $f(x) = \tan x$ we obtain

$$u_1' = W_1 = \begin{vmatrix} 0 & \cos x & \sin x \\ 0 & -\sin x & \cos x \\ \tan x & -\cos x & -\sin x \end{vmatrix} = \tan x$$

$$u_2' = W_2 = \begin{vmatrix} 1 & 0 & \sin x \\ 0 & 0 & \cos x \\ 0 & \tan x & -\sin x \end{vmatrix} = -\sin x$$

$$u_3' = W_3 = \begin{vmatrix} 1 & \cos x & 0 \\ 0 & -\sin x & 0 \\ 0 & -\cos x & \tan x \end{vmatrix} = -\sin x \tan x = \frac{\cos^2 x - 1}{\cos x} = \cos x - \sec x.$$

Then

$$u_1 = -\ln|\cos x|$$

$$u_2 = \cos x$$

$$u_3 = \sin x - \ln|\sec x + \tan x|$$

and

$$\begin{aligned} y &= c_1 + c_2 \cos x + c_3 \sin x - \ln|\cos x| + \cos^2 x \\ &\quad + \sin^2 x - \sin x \ln|\sec x + \tan x| \\ &= c_4 + c_2 \cos x + c_3 \sin x - \ln|\cos x| - \sin x \ln|\sec x + \tan x| \end{aligned}$$

for $-\infty < x < \infty$.

24. The auxiliary equation is $2m^3 - 6m^2 = 2m^2(m-3) = 0$, so $y_c = c_1 + c_2 x + c_3 e^{3x}$ and

$$W = \begin{vmatrix} 1 & x & e^{3x} \\ 0 & 1 & 3e^{3x} \\ 0 & 0 & 9e^{3x} \end{vmatrix} = 9e^{3x}.$$

Identifying $f(x) = x^2/2$ we obtain

$$u_1' = \frac{1}{9e^{3x}} W_1 = \frac{1}{9e^{3x}} \begin{vmatrix} 0 & x & e^{3x} \\ 0 & 1 & 3e^{3x} \\ x^2/2 & 0 & 9e^{3x} \end{vmatrix} = \frac{\frac{3}{2}x^3e^{3x} - \frac{1}{2}x^2e^{3x}}{9e^{3x}} = \frac{1}{6}x^3 - \frac{1}{18}x^2$$

$$u_2' = \frac{1}{9e^{3x}} W_2 = \frac{1}{9e^{3x}} \begin{vmatrix} 1 & 0 & e^{3x} \\ 0 & 0 & 3e^{3x} \\ 0 & x^2/2 & 9e^{3x} \end{vmatrix} = \frac{-\frac{3}{2}x^2e^{3x}}{9e^{3x}} = -\frac{1}{6}x^2$$

$$u_3' = \frac{1}{9e^{3x}} W_3 = \frac{1}{9e^{3x}} \begin{vmatrix} 1 & x & 0 \\ 0 & 1 & 0 \\ 0 & 0 & x^2/2 \end{vmatrix} = \frac{\frac{1}{2}x^2}{9e^{3x}} = \frac{1}{18}x^2e^{-3x}.$$

Then

$$u_1 = \frac{1}{24}x^4 - \frac{1}{54}x^3$$

$$u_2 = -\frac{1}{18}x^3$$

$$u_3 = -\frac{1}{54}x^2e^{-3x} - \frac{1}{81}xe^{-3x} - \frac{1}{243}e^{-3x}$$

and

$$\begin{aligned} y &= c_1 + c_2x + c_3e^{3x} + \frac{1}{24}x^4 - \frac{1}{54}x^3 - \frac{1}{18}x^4 - \frac{1}{54}x^2 - \frac{1}{81}x - \frac{1}{243} \\ &= c_4 + c_5x + c_3e^{3x} - \frac{1}{72}x^4 - \frac{1}{54}x^3 - \frac{1}{54}x^2 \end{aligned}$$

for $-\infty < x < \infty$.

27. The auxiliary equation is $m^2 + 2m - 8 = (m-2)(m+4) = 0$, so $y_c = c_1e^{2x} + c_2e^{-4x}$ and

$$W = \begin{vmatrix} e^{2x} & e^{-4x} \\ 2e^{2x} & -4e^{-4x} \end{vmatrix} = -6e^{-2x}.$$

Identifying $f(x) = 2e^{-2x} - e^{-x}$ we obtain

$$u_1' = \frac{1}{3}e^{-4x} - \frac{1}{6}e^{-3x}$$

$$u_2' = -\frac{1}{6}e^{3x} - \frac{1}{3}e^{2x}.$$

Then

$$u_1 = -\frac{1}{12}e^{-4x} + \frac{1}{18}e^{-3x}$$

$$u_2 = \frac{1}{18}e^{3x} - \frac{1}{6}e^{2x}.$$

Thus

$$y = c_1e^{2x} + c_2e^{-4x} - \frac{1}{12}e^{-2x} + \frac{1}{18}e^{-x} + \frac{1}{18}e^{-x} - \frac{1}{6}e^{-2x}$$
$$= c_1e^{2x} + c_2e^{-4x} - \frac{1}{4}e^{-2x} + \frac{1}{9}e^{-x}$$

and

$$y' = 2c_1e^{2x} - 4c_2e^{-4x} + \frac{1}{2}e^{-2x} - \frac{1}{9}e^{-x}.$$

The initial conditions imply

$$c_1 + c_2 - \frac{5}{36} = 1$$
$$2c_1 - 4c_2 + \frac{7}{18} = 0.$$

Thus $c_1 = 25/36$ and $c_2 = 4/9$, and

$$y = \frac{25}{36}e^{2x} + \frac{4}{9}e^{-4x} - \frac{1}{4}e^{-2x} + \frac{1}{9}e^{-x}.$$

30. Write the equation in the form

$$y'' + \frac{1}{x}y' + \frac{1}{x^2}y = \frac{\sec(\ln x)}{x^2}$$

and identify $f(x) = \sec(\ln x)/x^2$. From $y_1 = \cos(\ln x)$ and $y_2 = \sin(\ln x)$ we compute

$$W = \begin{vmatrix} \cos(\ln x) & \sin(\ln x) \\ -\dfrac{\sin(\ln x)}{x} & \dfrac{\cos(\ln x)}{x} \end{vmatrix} = \frac{1}{x}.$$

Now

$$u_1' = -\frac{\tan(\ln x)}{x} \quad \text{so} \quad u_1 = \ln|\cos(\ln x)|,$$

and

$$u_2' = \frac{1}{x} \quad \text{so} \quad u_2 = \ln x.$$

Thus, a particular solution is

$$y_p = \cos(\ln x)\ln|\cos(\ln x)| + (\ln x)\sin(\ln x).$$

Exercises 4.7

3. The auxiliary equation is $m^2 = 0$ so that $y = c_1 + c_2 \ln x$.

6. The auxiliary equation is $m^2 + 4m + 3 = (m+1)(m+3) = 0$ so that $y = c_1 x^{-1} + c_2 x^{-3}$.

9. The auxiliary equation is $25m^2 + 1 = 0$ so that $y = c_1 \cos\left(\frac{1}{5}\ln x\right) + c_2\left(\frac{1}{5}\ln x\right)$.

12. The auxiliary equation is $m^2 + 7m + 6 = (m+1)(m+6) = 0$ so that $y = c_1 x^{-1} + c_2 x^{-6}$.

15. The auxiliary equation is $3m^2 + 3m + 1 = 0$ so that $y = x^{-1/2}\left[c_1 \cos\left(\frac{\sqrt{3}}{6}\ln x\right) + c_2 \sin\left(\frac{\sqrt{3}}{6}\ln x\right)\right]$.

18. Assuming that $y = x^m$ and substituting into the differential equation we obtain

$$m(m-1)(m-2) + m - 1 = m^3 - 3m^2 + 3m - 1 = (m-1)^3 = 0.$$

Thus

$$y = c_1 x + c_2 x \ln x + c_3 x(\ln x)^2.$$

21. Assuming that $y = x^m$ and substituting into the differential equation we obtain

$$m(m-1)(m-2)(m-3) + 6m(m-1)(m-2) = m^4 - 7m^2 + 6m = m(m-1)(m-2)(m+3) = 0.$$

Thus

$$y = c_1 + c_2 x + c_3 x^2 + c_4 x^{-3}.$$

24. The auxiliary equation is $m^2 - 6m + 8 = (m-2)(m-4) = 0$, so that

$$y = c_1 x^2 + c_2 x^4 \quad \text{and} \quad y' = 2c_1 x + 4c_2 x^3.$$

The initial conditions imply

$$4c_1 + 16c_2 = 32$$

$$4c_1 + 32c_2 = 0.$$

Thus, $c_1 = 16$, $c_2 = -2$, and $y = 16x^2 - 2x^4$.

27. In this problem we use the substitution $t = -x$ since the initial conditions are on the interval $(-\infty, 0)$. Then

$$\frac{dy}{dt} = \frac{dy}{dx}\frac{dx}{dt} = -\frac{dy}{dx}$$

and

$$\frac{d^2y}{dt^2} = \frac{d}{dt}\left(\frac{dy}{dt}\right) = \frac{d}{dt}\left(-\frac{dy}{dx}\right) = -\frac{d}{dt}(y') = -\frac{dy'}{dx}\frac{dx}{dt} = -\frac{d^2y}{dx^2}\frac{dx}{dt} = \frac{d^2y}{dx^2},$$

so the differential equation and initial conditions become

$$4t^2\frac{d^2y}{dt^2} + y = 0; \quad y(t)\Big|_{t=1} = 2, \quad y'(t)\Big|_{t=1} = -4.$$

The auxiliary equation is $4m^2 - 4m + 1 = (2m-1)^2 = 0$, so that

$$y = c_1 t^{1/2} + c_2 t^{1/2} \ln t \quad \text{and} \quad y' = \frac{1}{2}c_1 t^{-1/2} + c_2\left(t^{-1/2} + \frac{1}{2}t^{-1/2}\ln t\right).$$

The initial conditions imply $c_1 = 2$ and $1 + c_2 = -4$. Thus

$$y = 2t^{1/2} - 5t^{1/2}\ln t = 2(-x)^{1/2} - 5(-x)^{1/2}\ln(-x), \quad x < 0.$$

30. The auxiliary equation is $m^2 - 5m = m(m-5) = 0$ so that $y_c = c_1 + c_2 x^5$ and

$$W(1, x^5) = \begin{vmatrix} 1 & x^5 \\ 0 & 5x^4 \end{vmatrix} = 5x^4.$$

Identifying $f(x) = x^3$ we obtain $u_1' = -\frac{1}{5}x^4$ and $u_2' = 1/5x$. Then $u_1 = -\frac{1}{25}x^5$, $u_2 = \frac{1}{5}\ln x$, and

$$y = c_1 + c_2 x^5 - \frac{1}{25}x^5 + \frac{1}{5}x^5 \ln x = c_1 + c_3 x^5 + \frac{1}{5}x^5 \ln x.$$

33. The auxiliary equation is $m^2 - 2m + 1 = (m-1)^2 = 0$ so that $y_c = c_1 x + c_2 x \ln x$ and

$$W(x, x\ln x) = \begin{vmatrix} x & x\ln x \\ 1 & 1 + \ln x \end{vmatrix} = x.$$

Identifying $f(x) = 2/x$ we obtain $u_1' = -2\ln x/x$ and $u_2' = 2/x$. Then $u_1 = -(\ln x)^2$, $u_2 = 2\ln x$, and

$$\begin{aligned} y &= c_1 x + c_2 x \ln x - x(\ln x)^2 + 2x(\ln x)^2 \\ &= c_1 x + c_2 x \ln x + x(\ln x)^2. \end{aligned}$$

In Problems 36-39 we use the following results: When $x = e^t$ or $t = \ln x$, then

$$\frac{dy}{dx} = \frac{1}{x}\frac{dy}{dt} \quad \text{and} \quad \frac{d^2y}{dx^2} = \frac{1}{x^2}\left[\frac{d^2y}{dt^2} - \frac{dy}{dt}\right].$$

36. Substituting into the differential equation we obtain

$$\frac{d^2y}{dt^2} - 5\frac{dy}{dt} + 6y = 2t.$$

The auxiliary equation is $m^2 - 5m + 6 = (m-2)(m-3) = 0$ so that $y_c = c_1 e^{2t} + c_2 e^{3t}$. Using undetermined coefficients we try $y_p = At + B$. This leads to $(-5A + 6B) + 6At = 2t$, so that $A = 1/3$, $B = 5/18$, and

$$y = c_1 e^{2t} + c_2 e^{3t} + \frac{1}{3}t + \frac{5}{18} = c_1 x^2 + c_2 x^3 + \frac{1}{3}\ln x + \frac{5}{18}.$$

39. Substituting into the differential equation we obtain

$$\frac{d^2y}{dt^2} + 8\frac{dy}{dt} - 20y = 5e^{-3t}.$$

The auxiliary equation is $m^2 + 8m - 20 = (m+10)(m-2) = 0$ so that $y_c = c_1e^{-10t} + c_2e^{2t}$. Using undetermined coefficients we try $y_p = Ae^{-3t}$. This leads to $-35Ae^{-3t} = 5e^{-3t}$, so that $A = -1/7$ and

$$y = c_1e^{-10t} + c_2e^{2t} - \frac{1}{7}e^{-3t} = c_1x^{-10} + c_2x^2 - \frac{1}{7}x^{-3}.$$

Exercises 4.8

3. From $Dx = -y + t$ and $Dy = x - t$ we obtain $y = t - Dx$, $Dy = 1 - D^2x$, and $(D^2+1)x = 1 + t$. Then

$$x = c_1\cos t + c_2\sin t + 1 + t$$

and

$$y = c_1\sin t - c_2\cos t + t - 1.$$

6. From $(D+1)x + (D-1)y = 2$ and $3x + (D+2)y = -1$ we obtain $x = -\frac{1}{3} - \frac{1}{3}(D+2)y$, $Dx = -\frac{1}{3}(D^2+2D)y$, and $(D^2+5)y = -7$. Then

$$y = c_1\cos\sqrt{5}\,t + c_2\sin\sqrt{5}\,t - \frac{7}{5}$$

and

$$x = \left(-\frac{2}{3}c_1 - \frac{\sqrt{5}}{3}c_2\right)\cos\sqrt{5}\,t + \left(\frac{\sqrt{5}}{3}c_1 - \frac{2}{3}c_2\right)\sin\sqrt{5}\,t + \frac{3}{5}.$$

9. From $Dx + D^2y = e^{3t}$ and $(D+1)x + (D-1)y = 4e^{3t}$ we obtain $D(D^2+1)x = 34e^{3t}$ and $D(D^2+1)y = -8e^{3t}$. Then

$$y = c_1 + c_2\sin t + c_3\cos t - \frac{4}{15}e^{3t}$$

and

$$x = c_4 + c_5\sin t + c_6\cos t + \frac{17}{15}e^{3t}.$$

Substituting into $(D+1)x + (D-1)y = 4e^{3t}$ gives

$$(c_4 - c_1) + (c_5 - c_6 - c_3 - c_2)\sin t + (c_6 + c_5 + c_2 - c_3)\cos t = 0$$

so that $c_4 = c_1$, $c_5 = c_3$, $c_6 = -c_2$, and

$$x = c_1 - c_2\cos t + c_3\sin t + \frac{17}{15}e^{3t}.$$

12. From $(2D^2-D-1)x-(2D+1)y=1$ and $(D-1)x+Dy=-1$ we obtain $(2D+1)(D-1)(D+1)x=-1$ and $(2D+1)(D+1)y=-2$. Then

$$x=c_1e^{-t/2}+c_2e^{-t}+c_3e^{t}+1$$

and

$$y=c_4e^{-t/2}+c_5e^{-t}-2.$$

Substituting into $(D-1)x+Dy=-1$ gives

$$\left(-\frac{3}{2}c_1-\frac{1}{2}c_4\right)e^{-t/2}+(-2c_2-c_5)e^{-t}=0$$

so that $c_4=-3c_1$, $c_5=-2c_2$, and

$$y=-3c_1e^{-t/2}-2c_2e^{-t}-2.$$

15. From $(D-1)x+(D^2+1)y=1$ and $(D^2-1)x+(D+1)y=2$ we obtain $D^2(D-1)(D+1)x=1$ and $D^2(D-1)(D+1)y=1$. Then

$$x=c_1+c_2t+c_3e^{t}+c_4e^{-t}-\frac{1}{2}t^2$$

and

$$y=c_5+c_6t+c_7e^{t}+c_8e^{-t}-\frac{1}{2}t^2.$$

Substituting into $(D-1)x+(D^2+1)y=1$ gives

$$(c_2-c_1-1+c_5)+(c_6-c_2-1)t+(2c_8-2c_4)e^{-t}+(2c_7)e^{t}=1$$

so that $c_6=c_2+1$, $c_8=c_4$, $c_7=0$, $c_5=c_1-c_2+2$, and

$$y=(c_1-c_2+2)+(c_2+1)t+c_4e^{-t}-\frac{1}{2}t^2.$$

18. From $Dx+z=e^t$, $(D-1)x+Dy+Dz=0$, and $x+2y+Dz=e^t$ we obtain $z=-Dx+e^t$, $Dz=-D^2x+e^t$, and the system $(-D^2+D-1)x+Dy=-e^t$ and $(-D^2+1)x+2y=0$. Then $y=\frac{1}{2}(D^2-1)x$, $Dy=\frac{1}{2}D(D^2-1)x$, and $(D-2)(D^2+1)x=-2e^t$ so that

$$x=c_1e^{2t}+c_2\cos t+c_3\sin t+e^t,$$

$$y=\frac{3}{2}c_1e^{2t}-c_2\cos t-c_3\sin t,$$

and

$$z=-2c_1e^{2t}-c_3\cos t+c_2\sin t.$$

21. From $2Dx+(D-1)y=t$ and $Dx+Dy=t^2$ we obtain $(D+1)y=2t^2-t$. Then

$$y=c_1e^{-t}+2t^2-5t+5$$

and $Dx = c_1e^{-t} + t^2 - 4t + 5$ so that

$$x = -c_1e^{-t} + c_2 + \frac{1}{3}t^3 - 2t^2 + 5t.$$

24. From $Dx - y = -1$ and $3x + (D-2)y = 0$ we obtain $x = -\frac{1}{3}(D-2)y$ so that $Dx = -\frac{1}{3}(D^2-2D)y$. Then $-\frac{1}{3}(D^2-2D)y = y-1$ and $(D^2-2D+3)y = 3$. Thus

$$y = e^t\left(c_1\cos\sqrt{2}\,t + c_2\sin\sqrt{2}\,t\right) + 1$$

and

$$x = \frac{1}{3}e^t\left[\left(c_1 - \sqrt{2}\,c_2\right)\cos\sqrt{2}\,t + \left(\sqrt{2}\,c_1 + c_2\right)\sin\sqrt{2}\,t\right] + \frac{2}{3}.$$

Using $x(0) = y(0) = 0$ we obtain

$$c_1 + 1 = 0$$

$$\frac{1}{3}\left(c_1 - \sqrt{2}\,c_2\right) + \frac{2}{3} = 0.$$

Thus $c_1 = -1$ and $c_2 = \sqrt{2}/2$. The solution of the initial value problem is

$$x = e^t\left(-\frac{2}{3}\cos\sqrt{2}\,t - \frac{\sqrt{2}}{6}\sin\sqrt{2}\,t\right) + \frac{2}{3}$$

$$y = e^t\left(-\cos\sqrt{2}\,t + \frac{\sqrt{2}}{2}\sin\sqrt{2}\,t\right) + 1.$$

Exercises 4.9

3. Let $u = y'$ so that $u' = y''$. The equation becomes $u' = -u - 1$ which is separable. Thus

$$\frac{du}{u^2+1} = -dx \implies \tan^{-1}u = -x + c_1 \implies y' = \tan(c_1 - x) \implies y = \ln|\cos(c_1 - x)| + c_2.$$

6. Let $u = y'$ so that $y'' = u\,\dfrac{du}{dy}$. The equation becomes $(y+1)u\,\dfrac{du}{dy} = u^2$. Separating variables we obtain

$$\frac{du}{u} = \frac{dy}{y+1} \implies \ln|u| = \ln|y+1| + \ln c_1 \implies u = c_1(y+1)$$

$$\implies \frac{dy}{dx} = c_1(y+1) \implies \frac{dy}{y+1} = c_1\,dx$$

$$\implies \ln|y+1| = c_1x + c_2 \implies y + 1 = c_3e^{c_1x}.$$

Exercises 4.9

9. Let $u = y'$ so that $y'' = u\dfrac{du}{dy}$. The equation becomes $u\dfrac{du}{dy} + yu = 0$. Separating variables we obtain

$$du = -y\,dy \implies u = -\frac{1}{2}y^2 + c_1 \implies y' = -\frac{1}{2}y^2 + c_1.$$

When $x = 0$, $y = 1$ and $y' = -1$ so $-1 = -\frac{1}{2} + c_1$ and $c_1 = -\frac{1}{2}$. Then

$$\frac{dy}{dx} = -\frac{1}{2}y^2 - \frac{1}{2} \implies \frac{dy}{y^2+1} = -\frac{1}{2}\,dx \implies \tan^{-1} y = -\frac{1}{2}x + c_2$$

$$\implies y = \tan\left(-\frac{1}{2}x + c_2\right).$$

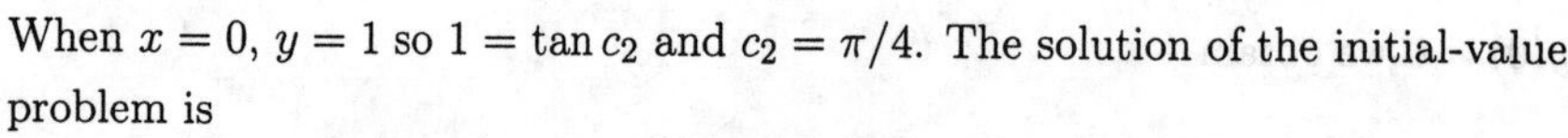

When $x = 0$, $y = 1$ so $1 = \tan c_2$ and $c_2 = \pi/4$. The solution of the initial-value problem is

$$y = \tan\left(\frac{\pi}{4} - \frac{1}{2}x\right), \qquad -\frac{\pi}{2} < x < \frac{3\pi}{2}.$$

12. Let $u = y'$ so that $u' = y''$. The equation becomes $u' - \dfrac{1}{x}u = u^2$, which is Bernoulli. Using the substitution $w = u^{-1}$ we obtain $\dfrac{dw}{dx} + \dfrac{1}{x}w = -1$. An integrating factor is x, so

$$\frac{d}{dx}[xw] = -x \implies w = -\frac{1}{2}x + \frac{1}{x}c \implies \frac{1}{u} = \frac{c_1 - x^2}{2x} \implies u = \frac{2x}{c_1 - x^2} \implies y = -\ln\left|c_1 - x^2\right| + c_2.$$

15. We look for a solution of the form

$$y(x) = y(0) + y'(0) + \frac{1}{2}y''(0) + \frac{1}{3!}y'''(0) + \frac{1}{4!}y^{(4)}(x) + \frac{1}{5!}y^{(5)}(x).$$

From $y''(x) = x^2 + y^2 - 2y'$ we compute

$$y'''(x) = 2x + 2yy' - 2y''$$

$$y^{(4)}(x) = 2 + 2(y')^2 + 2yy'' - 2y'''$$

$$y^{(5)}(x) = 6y'y'' + 2yy''' - 2y^{(4)}.$$

Using $y(0) = 1$ and $y'(0) = 1$ we find

$$y''(0) = -1, \quad y'''(0) = 4, \quad y^{(4)}(0) = -6, \quad y^{(5)}(0) = 14.$$

An approximate solution is

$$y(x) = 1 + x - \frac{1}{2}x^2 + \frac{2}{3}x^3 - \frac{1}{4}x^4 + \frac{7}{60}x^5.$$

18. Let $u = \dfrac{dx}{dt}$ so that $\dfrac{d^2x}{dt^2} = u\dfrac{du}{dx}$. The equation becomes $u\dfrac{du}{dx} = \dfrac{-k^2}{x^2}$. Separating variables we obtain

$$u\,du = -\frac{k^2}{x^2}\,dx \implies \frac{1}{2}u^2 = \frac{k^2}{x} + c \implies \frac{1}{2}v^2 = \frac{k^2}{x} + c.$$

When $t=0$, $x=x_0$ and $v=0$ so $0=\dfrac{k^2}{x_0}+c$ and $c=-\dfrac{k^2}{x_0}$. Then

$$\frac{1}{2}v^2=k^2\left(\frac{1}{x}-\frac{1}{x_0}\right) \quad \text{and} \quad \frac{dx}{dt}=-k\sqrt{2}\sqrt{\frac{x_0-x}{xx_0}}.$$

Separating variables we have

$$-\sqrt{\frac{xx_0}{x_0-x}}\,dx=k\sqrt{2}\,dt \implies t=-\frac{1}{k}\sqrt{\frac{x_0}{2}}\int\sqrt{\frac{x}{x_0-x}}\,dx.$$

Using *Mathematica* to integrate we obtain

$$\begin{aligned} t&=-\frac{1}{k}\sqrt{\frac{x_0}{2}}\left[-\sqrt{x(x_0-x)}-\frac{x_0}{2}\tan^{-1}\frac{(x_0-2x)}{2x}\sqrt{\frac{x}{x_0-x}}\right]\\ &=\frac{1}{k}\sqrt{\frac{x_0}{2}}\left[\sqrt{x(x_0-x)}+\frac{x_0}{2}\tan^{-1}\frac{x_0-2x}{2\sqrt{x(x_0-x)}}\right]. \end{aligned}$$

Chapter 4 Review Exercises

3. False; consider $f_1(x)=0$ and $f_2(x)=x$. These are linearly dependent even though x is not a multiple of 0. The statement would be true if it read "Two functions $f_1(x)$ and $f_2(x)$ are linearly independent on an interval if *neither* is a constant multiple of the other."

6. True

9. $A+Bxe^x$

12. Identifying $P(x)=-2-2/x$ we have $\int P\,dx=-2x-2\ln x$ and

$$y_2=e^x\int\frac{e^{2x+\ln x^2}}{e^{2x}}\,dx=e^x\int x^2\,dx=\frac{1}{3}x^3e^x.$$

15. From $m^3+10m^2+25m=0$ we obtain $m=0$, $m=-5$, and $m=-5$ so that

$$y=c_1+c_2e^{-5x}+c_3xe^{-5x}.$$

18. From $2m^4+3m^3+2m^2+6m-4=0$ we obtain $m=1/2$, $m=-2$, and $m=\pm\sqrt{2}\,i$ so that

$$y=c_1e^{x/2}+c_2e^{-2x}+c_3\cos\sqrt{2}\,x+c_4\sin\sqrt{2}\,x.$$

21. Applying D^4 to the differential equation we obtain $D^4(D^2-3D+5)=0$. Then

$$y=\underbrace{e^{3x/2}\left(c_1\cos\frac{\sqrt{11}}{2}x+c_2\sin\frac{\sqrt{11}}{2}x\right)}_{y_c}+c_3+c_4x+c_5x^2+c_6x^3$$

and $y_p = A + Bx + Cx^2 + Dx^3$. Substituting y_p into the differential equation yields

$$(5A - 3B + 2C) + (5B - 6C + 6D)x + (5C - 9D)x^2 + 5Dx^3 = -2x + 4x^3.$$

Equating coefficients gives $A = -222/625$, $B = 46/125$, $C = 36/25$, and $D = 4/5$. The general solution is

$$y = e^{3x/2}\left(c_1 \cos\frac{\sqrt{11}}{2}x + c_2 \sin\frac{\sqrt{11}}{2}x\right) - \frac{222}{625} + \frac{46}{125}x + \frac{36}{25}x^2 + \frac{4}{5}x^3.$$

24. Applying D to the differential equation we obtain $D(D^3 - D^2) = D^3(D-1) = 0$. Then

$$y = \underbrace{c_1 + c_2x + c_3e^x}_{y_c} + c_4x^2$$

and $y_p = Ax^2$. Substituting y_p into the differential equation yields $-2A = 6$. Equating coefficients gives $A = -3$. The general solution is

$$y = c_1 + c_2x + c_3e^x - 3x^2.$$

27. Let $u = y'$ so that $u' = y''$. The equation becomes $u\dfrac{du}{dx} = 4x$. Separating variables we obtain

$$u\,du = 4x\,dx \implies \frac{1}{2}u^2 = 2x^2 + c_1 \implies u^2 = 4x^2 + c_2.$$

When $x = 1$, $y' = u = 2$, so $4 = 4 + c_2$ and $c_2 = 0$. Then

$$u^2 = 4x^2 \implies \frac{dy}{dx} = 2x \quad \text{or} \quad \frac{dy}{dx} = -2x$$

$$\implies y = x^2 + c_3 \quad \text{or} \quad y = -x^2 + c_4.$$

When $x = 1$, $y = 5$, so $5 = 1 + c_3$ and $5 = -1 + c_4$. Thus $c_3 = 4$ and $c_4 = 6$. We have $y = x^2 + 4$ and $y = -x^2 + 6$. Note however that when $y = -x^2 + 6$, $y' = -2x$ and $y'(1) = -2 \neq 2$. Thus, the solution of the initial-value problem is $y = x^2 + 4$.

30. The auxiliary equation is $m^2 - 1 = 0$, so $y_c = c_1e^x + c_2e^{-x}$ and

$$W = \begin{vmatrix} e^x & e^{-x} \\ e^x & -e^{-x} \end{vmatrix} = -2.$$

Identifying $f(x) = 2e^x/(e^x + e^{-x})$ we obtain

$$u_1' = \frac{1}{e^x + e^{-x}} = \frac{e^x}{1 + e^{2x}}$$

$$u_2' = -\frac{e^{2x}}{e^x + e^{-x}} = -\frac{e^{3x}}{1 + e^{2x}} = -e^x + \frac{e^x}{1 + e^{2x}}.$$

Then $u_1 = \tan^{-1} e^x$, $u_2 = -e^x + \tan^{-1} e^x$, and

$$y = c_1e^x + c_2e^{-x} + e^x \tan^{-1} e^x - 1 + e^{-x} \tan^{-1} e^x.$$

33. The auxiliary equation is $2m^3 - 13m^2 + 24m - 9 = (2m-1)(m-3)^2 = 0$ so that

$$y_c = c_1 e^{x/2} + c_2 e^{3x} + c_3 x e^{3x}.$$

A particular solution is $y_p = -4$ and the general solution is

$$y = c_1 e^{x/2} + c_2 e^{3x} + c_3 x e^{3x} - 4.$$

Setting $y(0) = -4$, $y'(0) = 0$, and $y''(0) = \frac{5}{2}$ we obtain

$$c_1 + c_2 - 4 = -4$$

$$\frac{1}{2}c_1 + 3c_2 + c_3 = 0$$

$$\frac{1}{4}c_1 + 9c_2 + 6c_3 = \frac{5}{2}.$$

Solving this system we find $c_1 = \frac{2}{5}$, $c_2 = -\frac{2}{5}$, and $c_3 = 1$. Thus

$$y = \frac{2}{5}e^{x/2} - \frac{2}{5}e^{3x} + xe^{3x} - 4.$$

36. From $(D-2)x - y = t - 2$ and $-3x + (D-4)y = -4t$ we obtain $(D-1)(D-5)x = 9 - 8t$. Then

$$x = c_1 e^t + c_2 e^{5t} - \frac{8}{5}t - \frac{3}{25}$$

and

$$y = (D-2)x - t + 2 = -c_1 e^t + 3c_2 e^{5t} + \frac{16}{25} + \frac{11}{25}t.$$

5 Modeling with Higher-Order Differential Equations

Exercises 5.1

3. From $\frac{3}{4}x'' + 72x = 0$, $x(0) = -1/4$, and $x'(0) = 0$ we obtain $x = -\frac{1}{4}\cos 4\sqrt{6}\,t$.

6. From $50x'' + 200x = 0$, $x(0) = 0$, and $x'(0) = -10$ we obtain $x = -5\sin 2t$ and $x' = -10\cos 2t$.

9. From $\frac{1}{4}x'' + x = 0$, $x(0) = 1/2$, and $x'(0) = 3/2$ we obtain

$$x = \frac{1}{2}\cos 2t + \frac{3}{4}\sin 2t = \frac{\sqrt{13}}{4}\sin(2t + 0.588).$$

12. From $x' + 9x = 0$, $x(0) = -1$, and $x'(0) = -\sqrt{3}$ we obtain

$$x = -\cos 3t - \frac{\sqrt{3}}{3}\sin 3t = \frac{2}{\sqrt{3}}\sin\left(37 + \frac{4\pi}{3}\right)$$

and $x' = 2\sqrt{3}\cos(3t + 4\pi/3)$. If $x' = 3$ then $t = -7\pi/18 + 2n\pi/3$ and $t = -\pi/2 + 2n\pi/3$ for $n = 1$, 2, 3, ...

18. (a) below (b) from rest

21. From $\frac{1}{8}x'' + x' + 2x = 0$, $x(0) = -1$, and $x'(0) = 8$ we obtain $x = 4te^{-4t} - e^{-4t}$ and $x' = 8e^{-4t} - 16te^{-4t}$. If $x = 0$ then $t = 1/4$ second. If $x' = 0$ then $t = 1/2$ second and the extreme displacement is $x = e^{-2}$ feet.

24. (a) $x = \frac{1}{3}e^{-8t}\left(4e^{6t} - 1\right)$ is never zero; the extreme displacement is $x(0) = 1$ meter.

(b) $x = \frac{1}{3}e^{-8t}\left(5 - 2e^{6t}\right) = 0$ when $t = \frac{1}{6}\ln\frac{5}{2} \approx 0.153$ second; if $x' = \frac{4}{3}e^{-8t}\left(e^{6t} - 10\right) = 0$ then $t = \frac{1}{6}\ln 10 \approx 0.384$ second and the extreme displacement is $x = -0.232$ meter.

27. From $\frac{5}{16}x'' + \beta x' + 5x = 0$ we find that the roots of the auxiliary equation are $m = -\frac{8}{5}\beta \pm \frac{4}{5}\sqrt{4\beta^2 - 25}$.

(a) If $4\beta^2 - 25 > 0$ then $\beta > 5/2$.

(b) If $4\beta^2 - 25 = 0$ then $\beta = 5/2$.

(c) If $4\beta^2 - 25 < 0$ then $0 < \beta < 5/2$.

30. (a) If $x'' + 2x' + 5x = 12\cos 2t + 3\sin 2t$, $x(0) = -1$, and $x'(0) = 5$ then $x_c = e^{-t}(c_1\cos 2t + c_2\sin 2t)$ and $x_p = 3\sin 2t$ so that the equation of motion is

$$x = e^{-t}\cos 2t + 3\sin 2t.$$

(b)

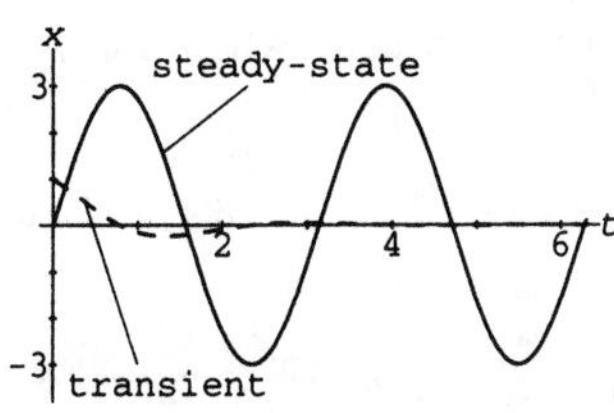

(c)

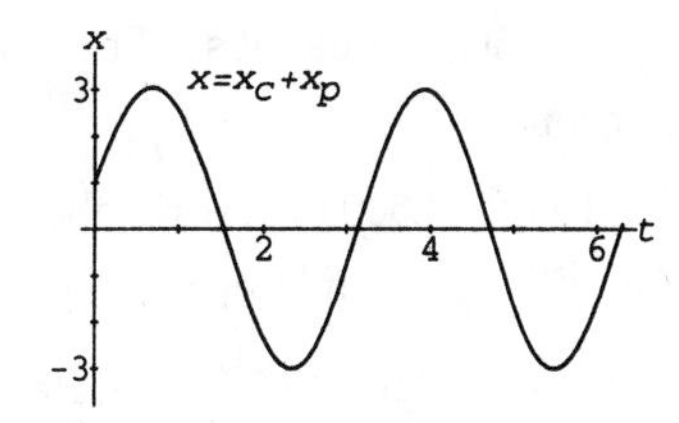

33. From $2x'' + 32x = 68e^{-2t}\cos 4t$, $x(0) = 0$, and $x'(0) = 0$ we obtain $x_c = c_1 \cos 4t + c_2 \sin 4t$ and $x_p = \frac{1}{2}e^{-2t}\cos 4t - 2e^{-2t}\sin 4t$ so that

$$x = -\frac{1}{2}\cos 4t + \frac{9}{4}\sin 4t + \frac{1}{2}e^{-2t}\cos 4t - 2e^{-2t}\sin 4t.$$

36. **(a)** From $100x'' + 1600x = 1600\sin 8t$, $x(0) = 0$, and $x'(0) = 0$ we obtain $x_c = c_1 \cos 4t + c_2 \sin 4t$ and $x_p = -\frac{1}{3}\sin 8t$ so that

$$x = \frac{2}{3}\sin 4t - \frac{1}{3}\sin 8t.$$

(b) If $x = \frac{1}{3}\sin 4t(2 - 2\cos 4t) = 0$ then $t = n\pi/4$ for $n = 0,\ 1,\ 2,\ \ldots$.

(c) If $x' = \frac{8}{3}\cos 4t - \frac{8}{3}\cos 8t = \frac{8}{3}(1 - \cos 4t)(1 + 2\cos 4t) = 0$ then $t = \pi/3 + n\pi/2$ and $t = \pi/6 + n\pi/2$ for $n = 0,\ 1,\ 2,\ \ldots$ at the extreme values. *Note*: There are many other values of t for which $x' = 0$.

(d) $x(\pi/6 + n\pi/2) = \sqrt{3}/2$ cm. and $x(\pi/3 + n\pi/2) = -\sqrt{3}/2$ cm.

(e)

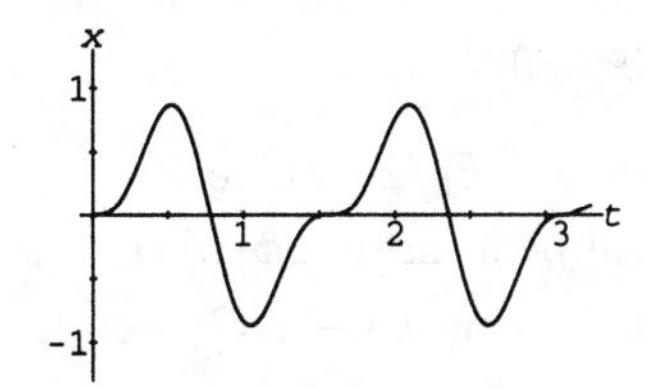

39. **(a)** From $x'' + \omega^2 x = F_0 \cos\gamma t$, $x(0) = 0$, and $x'(0) = 0$ we obtain $x_c = c_1 \cos\omega t + c_2 \sin\omega t$ and $x_p = (F_0 \cos\gamma t)/\left(\omega^2 - \gamma^2\right)$ so that

$$x = -\frac{F_0}{\omega^2 - \gamma^2}\cos\omega t + \frac{F_0}{\omega^2 - \gamma^2}\cos\gamma t.$$

(b) $\displaystyle \lim_{\gamma\to\omega} \frac{F_0}{\omega^2 - \gamma^2}(\cos\gamma t - \cos\omega t) = \lim_{\gamma\to\omega} \frac{-F_0 t \sin\gamma t}{-2\gamma} = \frac{F_0}{2\omega} t \sin\omega t.$

45. Solving $\frac{1}{20}q'' + 2q' + 100q = 0$ we obtain $q(t) = e^{-20t}(c_1 \cos 40t + c_2 \sin 40t)$. The initial conditions $q(0) = 5$ and $q'(0) = 0$ imply $c_1 = 5$ and $c_2 = 5/2$. Thus

$$q(t) = e^{-20t}\left(5\cos 40t + \frac{5}{2}\sin 40t\right) \approx \sqrt{25 + 25/4}\, e^{-20t}\sin(40t + 1.1071)$$

and $q(0.01) \approx 4.5676$ coulombs. The charge is zero for the first time when $40t + 0.4636 = \pi$ or $t \approx 0.0509$ second.

48. Solving $q'' + 100q' + 2500q = 30$ we obtain $q(t) = c_1 e^{-50t} + c_2 te^{-50t} + 0.012$. The initial conditions $q(0) = 0$ and $q'(0) = 2$ imply $c_1 = -0.012$ and $c_2 = 1.4$. Thus

$$q(t) = -0.012e^{-50t} + 1.4te^{-50t} + 0.012 \quad \text{and} \quad i(t) = 2e^{-50t} - 70te^{-50t}.$$

Solving $i(t) = 0$ we see that the maximum charge occurs when $t = 1/35$ and $q(1/35) \approx 0.01871$.

51. The differential equation is $\frac{1}{2}q'' + 20q' + 1000q = 100\sin t$. To use Example 11 in the text we identify $E_0 = 100$ and $\gamma = 60$. Then

$$X = L\gamma - \frac{1}{c\gamma} = \frac{1}{2}(60) - \frac{1}{0.001(60)} \approx 13.3333,$$

$$Z = \sqrt{X^2 + R^2} = \sqrt{X^2 + 400} \approx 24.0370,$$

and

$$\frac{E_0}{Z} = \frac{100}{Z} \approx 4.1603.$$

From Problem 50, then

$$i_p(t) \approx 4.1603(60t + \phi)$$

where $\sin\phi = -X/Z$ and $\cos\phi = R/Z$. Thus $\tan\phi = -X/R \approx -0.6667$ and ϕ is a fourth quadrant angle. Now $\phi \approx -0.5880$ and

$$i_p(t) \approx 4.1603(60t - 0.5880).$$

54. By Problem 50 the amplitude of the steady-state current is E_0/Z, where $Z = \sqrt{X^2 + R^2}$ and $X = L\gamma - 1/C\gamma$. Since E_0 is constant the amplitude will be a maximum when Z is a minimum. Since R is constant, Z will be a minimum when $X = 0$. Solving $L\gamma - 1/C\gamma = 0$ for γ we obtain $\gamma = 1/\sqrt{LC}$. The maximum amplitude will be E_0/R.

57. In an L-C series circuit there is no resistor, so the differential equation is

$$L\frac{d^2q}{dt^2} + \frac{1}{C}q = E(t).$$

Then $q(t) = c_1 \cos\left(t/\sqrt{LC}\right) + c_2 \sin\left(t/\sqrt{LC}\right) + q_p(t)$ where $q_p(t) = A\sin\gamma t + B\cos\gamma t$. Substituting $q_p(t)$ into the differential equation we find

$$\left(\frac{1}{C} - L\gamma^2\right) A\sin\gamma t + \left(\frac{1}{C} - L\gamma^2\right) B\cos\gamma t = E_0\cos\gamma t.$$

Equating coefficients we obtain $A = 0$ and $B = E_0C/(1 - LC\gamma^2)$. Thus, the charge is

$$q(t) = c_1 \cos\frac{1}{\sqrt{LC}}t + c_2 \sin\frac{1}{\sqrt{LC}}t + \frac{E_0C}{1 - LC\gamma^2}\cos\gamma t.$$

The initial conditions $q(0) = q_0$ and $q'(0) = i_0$ imply $c_1 = q_o - E_0C/(1 - LC\gamma^2)$ and $c_2 = i_0\sqrt{LC}$. The current is

$$i(t) = -\frac{c_1}{\sqrt{LC}}\sin\frac{1}{\sqrt{LC}}t + \frac{c_2}{\sqrt{LC}}\cos\frac{1}{\sqrt{LC}}t - \frac{E_0C\gamma}{1 - LC\gamma^2}\sin\gamma t$$

$$= i_0\cos\frac{1}{\sqrt{LC}}t - \frac{1}{\sqrt{LC}}\left(q_0 - \frac{E_0C}{1 - LC\gamma^2}\right)\sin\frac{1}{\sqrt{LC}}t - \frac{E_0C\gamma}{1 - LC\gamma^2}\sin\gamma t.$$

Exercises 5.2

3. (a) The general solution is

$$y(x) = c_1 + c_2x + c_3x^2 + c_4x^3 + \frac{w_0}{24EI}x^4.$$

The boundary conditions are $y(0) = 0$, $y'(0) = 0$, $y(L) = 0$, $y''(L) = 0$. The first two conditions give $c_1 = 0$ and $c_2 = 0$. The conditions at $x = L$ give the system

$$c_3L^2 + c_4L^3 + \frac{w_0}{24EI}L^4 = 0$$

$$2c_3 + 6c_4L + \frac{w_0}{2EI}L^2 = 0.$$

Solving, we obtain $c_3 = w_0L^2/16EI$ and $c_4 = -5w_0L/48EI$. The deflection is

$$y(x) = \frac{w_0}{48EI}(3L^2x^2 - 5Lx^3 + 2x^4).$$

(b)

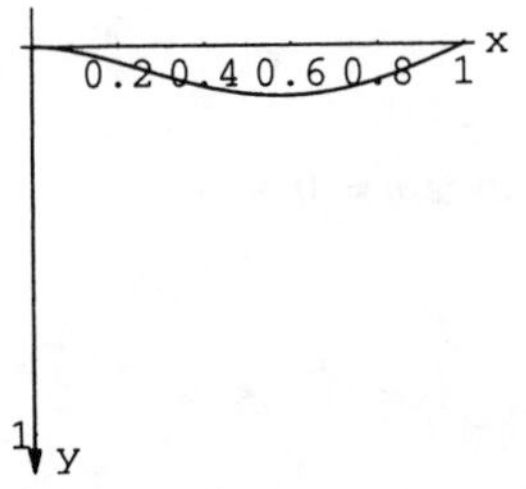

6. (a) $y_{\max} = y(L/2) = \dfrac{5w_0L^4}{384EI}$

(b) The maximum deflection of the beam in Example 1 is $y(L/2) = (w_0/24EI)L^4/16 = w_0L^4/384EI$, which is 1/5 of the maximum displacement of the beam in Problem 2.

9. For $\lambda \le 0$ the only solution of the boundary-value problem is $y = 0$. For $\lambda > 0$ we have

$$y = c_1\cos\sqrt{\lambda}\,x + c_2\sin\sqrt{\lambda}\,x.$$

Now $y(0) = 0$ implies $c_1 = 0$, so

$$y(\pi) = c_2 \sin \sqrt{\lambda}\,\pi = 0$$

gives

$$\sqrt{\lambda}\,\pi = n\pi \quad \text{or} \quad \lambda = n^2, \; n = 1, 2, 3, \ldots .$$

The eigenvalues n^2 correspond to the eigenfunctions $\sin nx$ for $n = 1, 2, 3, \ldots$.

12. For $\lambda \le 0$ the only solution of the boundary-value problem is $y = 0$. For $\lambda > 0$ we have

$$y = c_1 \cos \sqrt{\lambda}\,x + c_2 \sin \sqrt{\lambda}\,x.$$

Now $y(0) = 0$ implies $c_1 = 0$, so

$$y'\left(\frac{\pi}{2}\right) = c_2\sqrt{\lambda} \cos \sqrt{\lambda}\,\frac{\pi}{2} = 0$$

gives

$$\sqrt{\lambda}\,\frac{\pi}{2} = \frac{(2n-1)\pi}{2} \quad \text{or} \quad \lambda = (2n-1)^2, \; n = 1, 2, 3, \ldots .$$

The eigenvalues $(2n-1)^2$ correspond to the eigenfunctions $\sin(2n-1)x$.

15. The auxiliary equation has solutions

$$m = \frac{1}{2}\left(-2 \pm \sqrt{4 - 4(\lambda+1)}\right) = -1 \pm \sqrt{-\lambda}.$$

For $\lambda < 0$ we have

$$y = e^{-x}\left(c_1 \cosh \sqrt{-\lambda}\,x + c_2 \sinh \sqrt{-\lambda}\,x\right).$$

The boundary conditions imply

$$y(0) = c_1 = 0$$

$$y(5) = c_2 e^{-5} \sinh 5\sqrt{-\lambda} = 0$$

so $c_1 = c_2 = 0$ and the only solution of the boundary-value problem is $y = 0$.

For $\lambda = 0$ we have

$$y = c_1 e^{-x} + c_2 x e^{-x}$$

and the only solution of the boundary-value problem is $y = 0$.

For $\lambda > 0$ we have

$$y = e^{-x}\left(c_1 \cos \sqrt{\lambda}\,x + c_2 \sin \sqrt{\lambda}\,x\right).$$

Now $y(0) = 0$ implies $c_1 = 0$, so

$$y(5) = c_2 e^{-5} \sin 5\sqrt{\lambda} = 0$$

gives

$$5\sqrt{\lambda} = n\pi \quad \text{or} \quad \lambda = \frac{n^2\pi^2}{25}, \; n = 1, 2, 3, \ldots .$$

The eigenvalues $n^2\pi^2/25$ correspond to the eigenfunctions $e^{-x}\sin\frac{n\pi}{5}x$ for $n = 1, 2, 3, \ldots$.

18. For $\lambda = 0$ the only solution of the boundary-value problem is $y = 0$. For $\lambda \neq 0$ we have

$$y = c_1 \cos\lambda x + c_2 \sin\lambda x.$$

Now $y(0) = 0$ implies $c_1 = 0$, so

$$y'(3\pi) = c_2\lambda\cos 3\pi\lambda = 0$$

gives

$$3\pi\lambda = \frac{(2n-1)\pi}{2} \quad \text{or} \quad \lambda = \frac{2n-1}{6}, \; n = 1, 2, 3, \ldots.$$

The eigenvalues $(2n-1)/6$ correspond to the eigenfunctions $\sin\frac{2n-1}{6}x$ for $n = 1, 2, 3, \ldots$.

21. For $\lambda = 0$ the general solution is $y = c_1 + c_2 \ln x$. Now $y' = c_2/x$, so $y'(1) = c_2 = 0$ and $y = c_1$. Since $y'(e^2) = 0$ for any c_1 we see that $y(x) = 1$ is an eigenfunction corresponding to the eigenvalue $\lambda = 0$.

For $\lambda < 0$, $y = c_1 x^{-\sqrt{-\lambda}} + c_2 x^{\sqrt{-\lambda}}$. The initial conditions imply $c_1 = c_2 = 0$, so $y(x) = 0$.

For $\lambda > 0$, $y = c_1\cos(\sqrt{\lambda}\ln x) + c_2\sin(\sqrt{\lambda}\ln x)$. Now

$$y' = -c_1\frac{\sqrt{\lambda}}{x}\sin(\sqrt{\lambda}\ln x) + c_2\frac{\sqrt{\lambda}}{x}\cos(\sqrt{\lambda}\ln x),$$

and $y'(1) = c_2\sqrt{\lambda} = 0$ implies $c_2 = 0$. Finally, $y'(e^2) = -(c_1\sqrt{\lambda}/e^2)\sin(2\sqrt{\lambda}) = 0$ implies $\lambda = n^2\pi^2/4$ for $n = 1, 2, 3, \ldots$. The corresponding eigenfunctions are

$$y = \cos\left(\frac{n\pi}{2}\ln x\right).$$

24. **(a)** Since $\lambda_n = x_n^2$, there are no new eigenvalues when $x_n < 0$. For $\lambda = 0$, the differential equation $y'' = 0$ has general solution $y = c_1 x + c_2$. The boundary conditions imply $c_1 = c_2 = 0$, so $y = 0$.

(b) $\lambda_1 = 4.1159$, $\lambda_2 = 24.1393$, $\lambda_3 = 63.6591$, $\lambda_4 = 122.8892$.

27. The auxiliary equation is $m^2 + m = m(m+1) = 0$ so that $u(r) = c_1 r^{-1} + c_2$. The boundary conditions $u(a) = u_0$ and $u(b) = u_1$ yield the system $c_1 a^{-1} + c_2 = u_0$, $c_1 b^{-1} + c_2 = u_1$. Solving gives

$$c_1 = \left(\frac{u_0 - u_1}{b-a}\right)ab \quad \text{and} \quad c_2 = \frac{u_1 b - u_0 a}{b-a}.$$

Thus

$$u(r) = \left(\frac{u_0 - u_1}{b-a}\right)\frac{ab}{r} + \frac{u_1 b - u_0 a}{b-a}.$$

Exercises 5.3

3. The period corresponding to $x(0) = 1$, $x'(0) = 1$ is approximately 5.8. The second initial-value problem does not have a periodic solution.

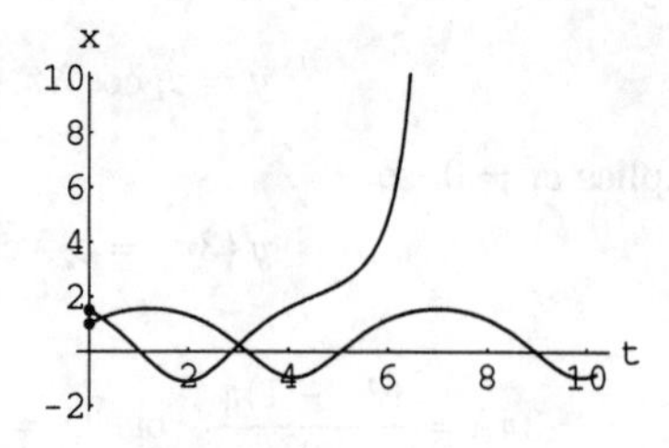

6. From the graphs we see that the interval is approximately $(-0.8, 1.1)$.

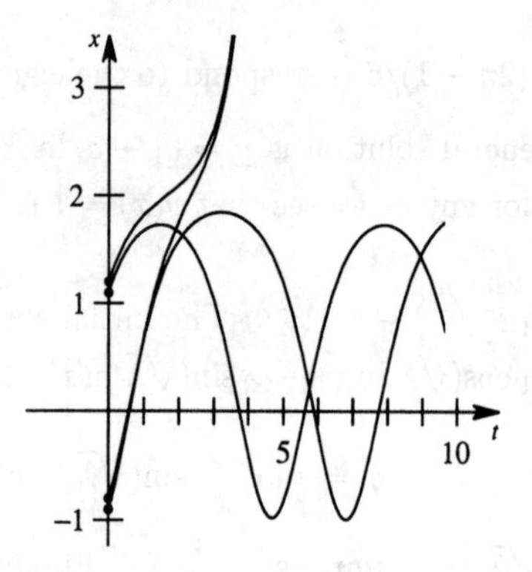

9. **(a)** This is a damped hard spring, so all solutions should be oscillatory with $x \to 0$ as $t \to \infty$.

(b)

12. **(a)**

k = -0.000471   k = -0.000472

The system appears to be oscillatory for $-0.000471 \le k_1 < 0$ and nonoscillatory for $k_1 \le -0.000472$.

(b)

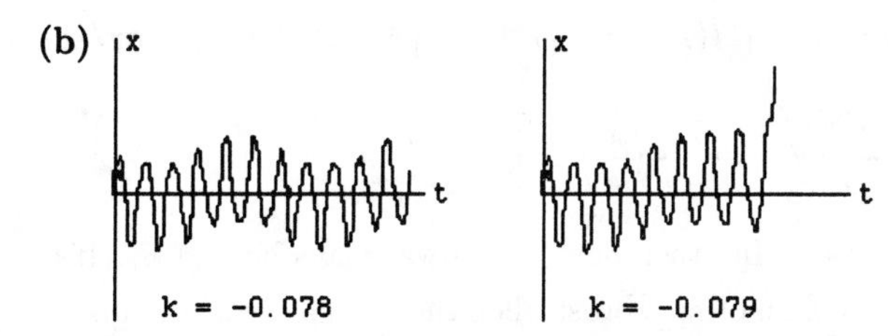

The system appears to be oscillatory for $-0.077 \leq k_1 < 0$ and nonoscillatory for $k_1 \leq 0.078$.

15. (a) Let (x, y) be the coordinates of S_2 on the curve C. The slope at (x, y) is then

$$dy/dx = (v_1 t - y)/(0 - x) = (y - v_1 t)/x \quad \text{or} \quad xy' - y = -v_1 t.$$

(b) Differentiating with respect to x gives

$$\begin{aligned} xy'' + y' - y' &= -v_1 \frac{dt}{dx} \\ xy'' &= -v_1 \frac{dt}{ds}\frac{ds}{dx} \\ xy'' &= -v_1 \frac{1}{v_2}(-\sqrt{1 + (y')^2}\,) \\ xy'' &= r\sqrt{1 + (y')^2}\,. \end{aligned}$$

Letting $u = y'$ and separating variables, we obtain

$$\begin{aligned} x\frac{du}{dx} &= r\sqrt{1 + u^2} \\ \frac{du}{\sqrt{1 + u^2}} &= \frac{r}{x}\,dx \\ \sinh^{-1} u &= r\ln x + \ln c = \ln(cx^r) \\ u &= \sinh(\ln cx^r) \\ \frac{dy}{dx} &= \frac{1}{2}\left(cx^r - \frac{1}{cx^r}\right). \end{aligned}$$

At $t = 0$, $dy/dx = 0$ and $x = a$, so $0 = ca^r - 1/ca^r$. Thus $c = 1/a^r$ and

$$\frac{dy}{dx} = \frac{1}{2}\left[\left(\frac{x}{a}\right)^r - \left(\frac{a}{x}\right)^r\right] = \frac{1}{2}\left[\left(\frac{x}{a}\right)^r - \left(\frac{x}{a}\right)^{-r}\right].$$

If $r > 1$ or $r < 1$, integrating gives

$$y = \frac{a}{2}\left[\frac{1}{1 + r}\left(\frac{x}{a}\right)^{1+r} - \frac{1}{1 - r}\left(\frac{x}{a}\right)^{1-r}\right] + c_1.$$

When $t=0$, $y=0$ and $x=a$, so $0=(a/2)[1/(1+r)-1/(1-r)]+c_1$. Thus $c_1=ar/(1-r^2)$ and

$$y=\frac{a}{2}\left[\frac{1}{1+r}\left(\frac{x}{a}\right)^{1+r}-\frac{1}{1-r}\left(\frac{x}{a}\right)^{1-r}\right]+\frac{ar}{1-r^2}.$$

(c) If $r>1$, $v_1>v_2$ and $y\to\infty$ as $x\to 0^+$. In other words, S_2 always lags behind S_1. If $r<1$, $v_1<v_2$ and $y=ar/(1-r^2)$ when $x=0$. In other words, when the submarine's speed is greater than the ship's, their paths will intersect at the point $(0,ar/(1-r^2))$.
If $r=1$, integration gives

$$y=\frac{1}{2}\left[\frac{x^2}{2a}-\frac{1}{a}\ln x\right]+c_2.$$

When $t=0$, $y=0$ and $x=a$, so $0=(1/2)[a/2-(1/a)\ln a]+c_2$. Thus $c_2=-(1/2)[a/2-(1/a)\ln a]$ and

$$y=\frac{1}{2}\left[\frac{x^2}{2a}-\frac{1}{a}\ln x\right]-\frac{1}{2}\left[\frac{a}{2}-\frac{1}{a}\ln a\right]=\frac{1}{2}\left[\frac{1}{2a}(x^2-a^2)+\frac{1}{a}\ln\frac{a}{x}\right].$$

Since $y\to\infty$ as $x\to 0^+$, S_2 will never catch up with S_1.

Chapter 5 Review Exercises

3. 5/4 m., since $x=-\cos 4t+\frac{3}{4}\sin 4t$.

6. False

9. 9/2, since $x=c_1\cos\sqrt{2k}\,t+c_2\sin\sqrt{2k}\,t$.

12. From $x''+\beta x'+64x=0$ we see that oscillatory motion results if $\beta^2-256<0$ or $0\le|\beta|<16$.

15. Writing $\frac{1}{8}x''+\frac{8}{3}x=\cos\gamma t+\sin\gamma t$ in the form $x''+\frac{64}{3}x=8\cos\gamma t+8\sin\gamma t$ we identify $\lambda=0$ and $\omega^2=64/3$. The system is in a state of pure resonance when $\gamma=\omega=\sqrt{64/3}=8/\sqrt{3}$.

18. (a) Let k be the effective spring constant and x_1 and x_2 the elongation of springs k_1 and k_2. The restoring forces satisfy $k_1x_1=k_2x_2$ so $x_2=(k_1/k_2)x_1$. From $k(x_1+x_2)=k_1x_1$ we have

$$k\left(x_1+\frac{k_1}{k_2}x_2\right)=k_1x_1$$

$$k\left(\frac{k_2+k_1}{k_2}\right)=k_1$$

$$k=\frac{k_1k_2}{k_1+k_2}$$

$$\frac{1}{k}=\frac{1}{k_1}+\frac{1}{k_2}.$$

(b) From $k_1 = 2W$ and $k_2 = 4W$ we find $1/k = 1/2W + 1/4W = 3/4W$. Then $k = 4W/3 = 4mg/3$. The differential equation $mx'' + kx = 0$ then becomes $x'' + (4g/3)x = 0$. The solution is

$$x(t) = c_1 \cos 2\sqrt{\frac{g}{3}}\,t + c_2 \sin 2\sqrt{\frac{g}{3}}\,t.$$

The initial conditions $x(0) = 1$ and $x'(0) = 2/3$ imply $c_1 = 1$ and $c_2 = 1/\sqrt{3g}$.

(c) To compute the maximum speed of the weight we compute

$$x'(t) = 2\sqrt{\frac{g}{3}} \sin 2\sqrt{\frac{g}{3}}\,t + \frac{2}{3} \cos 2\sqrt{\frac{g}{3}}\,t \quad \text{and} \quad |x'(t)| = \sqrt{4\frac{g}{3} + \frac{4}{9}} = \frac{2}{3}\sqrt{3g+1}.$$

21. For $\lambda > 0$ the general solution is $y = c_1 \cos\sqrt{\lambda}\,x + c_2 \sin\sqrt{\lambda}\,x$. Now $y(0) = c_1$ and $y(2\pi) = c_1 \cos 2\pi\sqrt{\lambda} + c_2 \sin 2\pi\sqrt{\lambda}$, so the condition $y(0) = y(2\pi)$ implies

$$c_1 = c_1 \cos 2\pi\sqrt{\lambda} + c_2 \sin 2\pi\sqrt{\lambda}$$

which is true when $\sqrt{\lambda} = n$ or $\lambda = n^2$ for $n = 1, 2, 3, \ldots$. Since $y' = -\sqrt{\lambda}\,c_1 \sin\sqrt{\lambda}\,x + \sqrt{\lambda}\,c_2 \cos\sqrt{\lambda}\,x = -nc_1 \sin nx + nc_2 \cos nx$, we see that $y'(0) = nc_2 = y'(2\pi)$ for $n = 1, 2, 3, \ldots$. Thus, the eigenvalues are n^2 for $n = 1, 2, 3, \ldots$, with corresponding eigenfunctions $\cos nx$ and $\sin nx$. When $\lambda = 0$, the general solution is $y = c_1 x + c_2$ and the corresponding eigenfunction is $y = 1$. For $\lambda < 0$ the general solution is $y = c_1 \cosh\sqrt{-\lambda}\,x + c_2 \sinh\sqrt{-\lambda}\,x$. In this case $y(0) = c_1$ and $y(2\pi) = c_1 \cosh 2\pi\sqrt{-\lambda} + c_2 \sinh 2\pi\sqrt{-\lambda}$, so $y(0) = y(2\pi)$ can only be valid for $\lambda = 0$. Thus, there are no eigenvalues corresponding to $\lambda < 0$.

6 Series Solutions of Linear Equations

Exercises 6.1

3. $\displaystyle\lim_{n\to\infty}\left|\frac{a_{n+1}}{a_n}\right| = \lim_{n\to\infty}\left|\frac{2^{n+1}x^{n+1}/(n+1)}{2^n x^n/n}\right| = \lim_{n\to\infty}\frac{2n}{n+1}|x| = 2|x|$

The series is absolutely convergent for $2|x| < 1$ or $|x| < 1/2$. At $x = -1/2$, the series $\displaystyle\sum_{k=1}^{\infty}\frac{(-1)^k}{k}$ converges by the alternating series test. At $x = 1/2$, the series $\displaystyle\sum_{k=1}^{\infty}\frac{1}{k}$ is the harmonic series which diverges. Thus, the given series converges on $[-1/2, 1/2)$.

6. $\displaystyle\lim_{n\to\infty}\left|\frac{a_{n+1}}{a_n}\right| = \lim_{n\to\infty}\left|\frac{(x+7)^{n+1}/\sqrt{n+1}}{(x+7)^n\sqrt{n}}\right| = \lim_{n\to\infty}\sqrt{\frac{n}{n+1}}\,|x+7| = |x+7|$

The series is absolutely convergent for $|x+7| < 1$ or on $(-8, 6)$. At $x = -8$, the series $\displaystyle\sum_{n=1}^{\infty}\frac{(-1)^n}{\sqrt{n}}$ converges by the alternating series test. At $x = -6$, the series $\displaystyle\sum_{n=1}^{\infty}\frac{1}{\sqrt{n}}$ is a divergent p-series. Thus, the given series converges on $[-8, -6)$.

9. $\displaystyle\lim_{n\to\infty}\left|\frac{a_{n+1}}{a_n}\right| = \lim_{n\to\infty}\left|\frac{(n+1)!2^{n+1}x^{n+1}}{n!2^n x^n}\right| = \lim_{n\to\infty} 2(n+1)|x| = \infty,\ \ x \neq 0$

The series converges only at $x = 0$.

12. $\displaystyle e^{-x}\cos x = \left(1 - x + \frac{x^2}{2} - \frac{x^3}{6} + \frac{x^4}{24} - \cdots\right)\left(1 - \frac{x^2}{2} + \frac{x^4}{24} - \cdots\right) = 1 - x + \frac{x^3}{3} - \frac{x^4}{6} + \cdots$

15. Separating variables we obtain

$$\frac{dy}{y} = -dx \implies \ln|y| = -x + c \implies y = c_1 e^{-x}.$$

Substituting $y = \sum_{n=0}^{\infty} c_n x^n$ into the differential equation leads to

$$y' + y = \underbrace{\sum_{n=1}^{\infty} nc_n x^{n-1}}_{k=n-1} + \underbrace{\sum_{n=0}^{\infty} c_n x^n}_{k=n} = \sum_{k=0}^{\infty}(k+1)c_{k+1}x^k + \sum_{k=0}^{\infty} c_k x^k = \sum_{k=0}^{\infty}[(k+1)c_{k+1} + c_k]x^k = 0.$$

Thus

$$(k+1)c_{k+1} + c_k = 0$$

and

$$c_{k+1} = -\frac{1}{k+1}c_k, \quad k = 0, 1, 2, \ldots .$$

Iterating we find

$$c_1 = -c_0$$

$$c_2 = -\frac{1}{2}c_1 = \frac{1}{2}c_0$$

$$c_3 = -\frac{1}{3}c_2 = -\frac{1}{6}c_0$$

$$c_4 = -\frac{1}{4}c_3 = \frac{1}{24}c_0$$

and so on. Therefore

$$y = c_0 - c_0x + \frac{1}{2}c_0x^2 - \frac{1}{6}c_0x^3 + \frac{1}{24}c_0x^4 - \cdots = c_0\left[1 - x + \frac{1}{2}x^2 - \frac{1}{6}x^3 + \frac{1}{24}x^4 - \cdots\right]$$

$$= c_0\sum_{n=0}^{\infty}\frac{1}{n!}(-x)^n = c_0e^{-x}.$$

18. Separating variables we obtain

$$\frac{dy}{y} = -x^3dx \implies \ln|y| = -\frac{1}{4}x^4 + c \implies y = c_1e^{-x^4/4}.$$

Substituting $y = \sum_{n=0}^{\infty} c_nx^n$ into the differential equation leads to

$$y' + x^3y = \underbrace{\sum_{n=1}^{\infty} nc_nx^{n-1}}_{k=n-4} + \underbrace{\sum_{n=0}^{\infty} c_nx^{n+3}}_{k=n} = \sum_{k=-3}^{\infty}(k+4)c_{k+4}x^{k+3} - \sum_{k=0}^{\infty} c_kx^{k+3}$$

$$= c_1 + 2c_2x + 3c_3x^2 + \sum_{k=0}^{\infty}[(k+4)c_{k+4} + c_k]x^{k+2} = 0.$$

Thus

$$c_1 = c_2 = c_3 = 0,$$

$$(k+4)c_{k+4} + c_k = 0,$$

and

$$c_{k+4} = -\frac{1}{k+4}c_k, \quad k = 0, 1, 2, \ldots .$$

Iterating we find

$$c_4 = -\frac{1}{4}c_0$$

$$c_5 = c_6 = c_7 = 0$$

$$c_8 = -\frac{1}{8}c_4 = \frac{1}{2}\cdot\frac{1}{4^2}c_0$$

$$c_9 = c_{10} = c_{11} = 0$$

$$c_{12} = -\frac{1}{12}c_8 = -\frac{1}{2\cdot 3}\cdot\frac{1}{4^3}c_0$$

and so on. Therefore

$$y = c_0 - \frac{1}{4}c_0x^4 + \frac{1}{2}\cdot\frac{1}{4^2}c_0x^8 - \frac{1}{2\cdot 3}\cdot\frac{1}{4^3}c_0x^{12} + \cdots$$

$$= c_0\left[1 - \frac{x^4}{4} + \frac{1}{2}\left(\frac{x^4}{4}\right)^2 - \frac{1}{2\cdot 3}\left(\frac{x^4}{4}\right)^3 + \cdots\right] = c_0\sum_{n=0}^{\infty}\frac{1}{n!}\left(\frac{-x^4}{4}\right)^n = c_0e^{-x^4/4}.$$

21. The auxiliary equation is $m^2+1=0$, so $y = c_1\cos x + c_2\sin x$. Substituting $y = \sum_{n=0}^{\infty} c_nx^n$ into the differential equation leads to

$$y'' + y = \underbrace{\sum_{n=2}^{\infty} n(n-1)c_nx^{n-2}}_{k=n-2} + \underbrace{\sum_{n=0}^{\infty} c_nx^n}_{k=n} = \sum_{k=0}^{\infty}(k+2)(k+1)c_{k+2}x^k + \sum_{k=0}^{\infty} c_kx^k$$

$$= \sum_{k=0}^{\infty}[(k+2)(k+1)c_{k+2} + c_k]x^k = 0.$$

Thus

$$(k+2)(k+1)c_{k+2} + c_k = 0$$

and

$$c_{k+2} = -\frac{1}{(k+2)(k+1)}c_k, \quad k = 0, 1, 2, \ldots .$$

Iterating we find

$$c_2 = -\frac{1}{2}c_0$$

$$c_3 = -\frac{1}{3\cdot 2}c_1$$

$$c_4 = -\frac{1}{4\cdot 3}c_2 = \frac{1}{4\cdot 3\cdot 2}c_0$$

$$c_5 = -\frac{1}{5\cdot 4}c_3 = \frac{1}{5\cdot 4\cdot 3\cdot 2}c_1$$

$$c_6 = -\frac{1}{6\cdot 5}c_4 = -\frac{1}{6!}c_0$$

$$c_7 = -\frac{1}{7\cdot 6}c_5 = -\frac{1}{7!}c_1$$

and so on. Therefore

$$y = c_0 + c_1x - \frac{1}{2}c_0x^2 - \frac{1}{3!}c_1x^3 + \frac{1}{4!}c_0x^4 + \frac{1}{5!}c_1x^5 - \cdots$$

$$= c_0\left[1 - \frac{1}{2}x^2 + \frac{1}{4!}x^4 - \cdots\right] + c_1\left[1 - \frac{1}{3!}x^3 + \frac{1}{5!}x^5 - \cdots\right]$$

$$= c_0\sum_{n=0}^{\infty}\frac{(-1)^nx^{2n}}{(2n)!} + c_1\sum_{n=0}^{\infty}\frac{(-1)^nx^{2n+1}}{(2n+1)!} = c_0\cos x + c_1\sin x.$$

24. The auxiliary equation is $2m^2 + m = m(2m+1) = 0$, so $y = c_1 + c_2 e^{-x/2}$. Substituting $y = \sum_{n=0}^{\infty} c_n x^n$ into the differential equation leads to

$$2y'' + y' = 2\underbrace{\sum_{n=2}^{\infty} n(n-1)c_n x^{n-2}}_{k=n-2} + \underbrace{\sum_{n=1}^{\infty} nc_n x^{n-1}}_{k=n-1}$$

$$= 2\sum_{k=0}^{\infty}(k+2)(k+1)c_{k+2}x^k + \sum_{k=0}^{\infty}(k+1)c_{k+1}x^k$$

$$= \sum_{k=0}^{\infty}[2(k+2)(k+1)c_{k+2} + (k+1)c_{k+1}]x^k = 0.$$

Thus

$$2(k+2)(k+1)c_{k+2} + (k+1)c_{k+1} = 0$$

and

$$c_{k+2} = -\frac{1}{2(k+2)}c_{k+1}, \quad k = 0, 1, 2, \ldots .$$

Iterating we find

$$c_2 = -\frac{1}{2}\frac{1}{2}c_1$$

$$c_3 = -\frac{1}{2}\frac{1}{3}c_2 = \frac{1}{2^2}\frac{1}{3\cdot 2}c_1$$

$$c_4 = -\frac{1}{2}\frac{1}{4}c_3 = \frac{1}{2^3}\frac{1}{4!}c_1$$

and so on. Therefore

$$y = c_0 + c_1 x - \frac{1}{2}\frac{1}{2}c_1 x^2 + \frac{1}{2^2 3!}c_1 x^3 - \frac{1}{2^3 4!}c_1 x^4 + \cdots$$

$$\boxed{\begin{aligned} c_0 &= C_0 - 2c_1 \\ c_1 &= -\tfrac{1}{2}C_1 \end{aligned}}$$

$$= C_0 + \left[C_1 - \frac{1}{2}C_1 x + \frac{1}{2}\frac{1}{2}\frac{1}{2}C_1 x^2 - \frac{1}{2^2 2\cdot 3!}\frac{1}{2}C_1 x^3 + \cdots\right]$$

$$= C_0 + C_1\left[1 - \frac{x}{2} + \frac{1}{2}\left(\frac{x}{2}\right)^2 - \frac{1}{3!}\left(\frac{x}{3}\right)^3 + \cdots\right]$$

$$= C_0 + C_1\sum_{n=0}^{\infty}\frac{(-1)^n}{n!}\left(\frac{x}{2}\right)^n = C_0 + C_1\sum_{n=0}^{\infty}\frac{1}{n!}\left(-\frac{x}{n}\right)^n = C_0 + C_1 e^{-x/2}.$$

Exercises 6.2

3. Substituting $y = \sum_{n=0}^{\infty} c_n x^n$ into the differential equation we have

$$y'' - 2xy' + y = \underbrace{\sum_{n=2}^{\infty} n(n-1)c_n x^{n-2}}_{k=n-2} - 2\underbrace{\sum_{n=1}^{\infty} nc_n x^n}_{k=n} + \underbrace{\sum_{n=0}^{\infty} c_n x^n}_{k=n}$$

$$= \sum_{k=0}^{\infty}(k+2)(k+1)c_{k+2}x^k - 2\sum_{k=1}^{\infty} kc_k x^k + \sum_{k=0}^{\infty} c_k x^k$$

$$= 2c_2 + c_0 + \sum_{k=1}^{\infty}[(k+2)(k+1)c_{k+2} - (2k-1)c_k]x^k = 0.$$

Thus

$$2c_2 + c_0 = 0$$

$$(k+2)(k+1)c_{k+2} - (2k-1)c_k = 0$$

and

$$c_2 = -\frac{1}{2}c_0$$

$$c_{k+2} = \frac{2k-1}{(k+2)(k+1)}c_k, \quad k = 1, 2, 3, \ldots.$$

Choosing $c_0 = 1$ and $c_1 = 0$ we find

$$c_2 = -\frac{1}{2}$$

$$c_3 = c_5 = c_7 = \cdots = 0$$

$$c_4 = -\frac{1}{8}$$

$$c_6 = -\frac{7}{336}$$

and so on. For $c_0 = 0$ and $c_1 = 1$ we obtain

$$c_2 = c_4 = c_6 = \cdots = 0$$

$$c_3 = \frac{1}{6}$$

$$c_5 = \frac{1}{24}$$

$$c_7 = \frac{1}{112}$$

and so on. Thus, two solutions are

$$y_1 = 1 - \frac{1}{2}x^2 - \frac{1}{8}x^4 - \frac{7}{336}x^6 - \cdots \quad \text{and} \quad y_2 = x + \frac{1}{6}x^3 + \frac{1}{24}x^5 + \frac{1}{112}x^7 + \cdots.$$

6. Substituting $y = \sum_{n=0}^{\infty} c_n x^n$ into the differential equation we have

$$\begin{aligned} y'' + 2xy' + 2y &= \underbrace{\sum_{n=2}^{\infty} n(n-1)c_n x^{n-2}}_{k=n-2} + \underbrace{2\sum_{n=1}^{\infty} nc_n x^n}_{k=n} + \underbrace{2\sum_{n=0}^{\infty} c_n x^n}_{k=n} \\ &= \sum_{k=0}^{\infty}(k+2)(k+1)c_{k+2}x^k + 2\sum_{k=1}^{\infty} kc_k x^k + 2\sum_{k=0}^{\infty} c_k x^k \\ &= 2c_2 + 2c_0 + \sum_{k=1}^{\infty}[(k+2)(k+1)c_{k+2} + 2(k+1)c_k]x^k = 0. \end{aligned}$$

Thus

$$2c_2 + 2c_0 = 0$$

$$(k+2)(k+1)c_{k+2} + 2(k+1)c_k = 0$$

and

$$c_2 = -c_0$$

$$c_{k+2} = -\frac{2}{k+2}c_k, \quad k = 1, 2, 3, \ldots.$$

Choosing $c_0 = 1$ and $c_1 = 0$ we find

$$c_2 = -1$$

$$c_3 = c_5 = c_7 = \cdots = 0$$

$$c_4 = \frac{1}{2}$$

$$c_6 = -\frac{1}{6}$$

and so on. For $c_0 = 0$ and $c_1 = 1$ we obtain

$$c_2 = c_4 = c_6 = \cdots = 0$$

$$c_3 = -\frac{2}{3}$$

$$c_5 = \frac{4}{15}$$

$$c_7 = -\frac{8}{105}$$

and so on. Thus, two solutions are

$$y_1 = 1 - x^2 + \frac{1}{2}x^4 - \frac{1}{6}x^6 + \cdots \quad \text{and} \quad y_2 = x - \frac{2}{3}x^3 + \frac{4}{15}x^5 - \frac{8}{105}x^7 + \cdots .$$

9. Substituting $y = \sum_{n=0}^{\infty} c_n x^n$ into the differential equation we have

$$\left(x^2 - 1\right)y'' + 4xy' + 2y = \underbrace{\sum_{n=2}^{\infty} n(n-1)c_n x^n}_{k=n} - \underbrace{\sum_{n=2}^{\infty} n(n-1)c_n x^{n-2}}_{k=n-2} + 4\underbrace{\sum_{n=1}^{\infty} nc_n x^n}_{k=n} + 2\underbrace{\sum_{n=0}^{\infty} c_n x^n}_{k=n}$$

$$= \sum_{k=2}^{\infty} k(k-1)c_k x^k - \sum_{k=0}^{\infty}(k+2)(k+1)c_{k+2}x^k + 4\sum_{k=1}^{\infty} kc_k x^k + 2\sum_{k=0}^{\infty} c_k x^k$$

$$= -2c_2 + 2c_0 + (-6c_3 + 6c_1)x + \sum_{k=2}^{\infty}\left[\left(k^2 - k + 4k + 2\right)c_k - (k+2)(k+1)c_{k+2}\right]x^k = 0.$$

Thus

$$-2c_2 + 2c_0 = 0$$

$$-6c_3 + 6c_1 = 0$$

$$\left(k^2 + 3k + 2\right)c_k - (k+2)(k+1)c_{k+2} = 0$$

and

$$c_2 = c_0$$

$$c_3 = c_1$$

$$c_{k+2} = c_k, \quad k = 2, 3, 4, \ldots .$$

Choosing $c_0 = 1$ and $c_1 = 0$ we find

$$c_2 = 1$$

$$c_3 = c_5 = c_7 = \cdots = 0$$

$$c_4 = c_6 = c_8 = \cdots = 1.$$

For $c_0 = 0$ and $c_1 = 1$ we obtain

$$c_2 = c_4 = c_6 = \cdots = 0$$

$$c_3 = c_5 = c_7 = \cdots = 1.$$

Thus, two solutions are

$$y_1 = 1 + x^2 + x^4 + \cdots \quad \text{and} \quad y_2 = x + x^3 + x^5 + \cdots .$$

12. Substituting $y=\sum_{n=0}^{\infty}c_nx^n$ into the differential equation we have

$$\left(x^2-1\right)y''+xy'-y=\underbrace{\sum_{n=2}^{\infty}n(n-1)c_nx^n}_{k=n}-\underbrace{\sum_{n=2}^{\infty}n(n-1)c_nx^{n-2}}_{k=n-2}+\underbrace{\sum_{n=1}^{\infty}nc_nx^n}_{k=n}-\underbrace{\sum_{n=0}^{\infty}c_nx^n}_{k=n}$$

$$=\sum_{k=2}^{\infty}k(k-1)c_kx^k-\sum_{k=0}^{\infty}(k+2)(k+1)c_{k+2}x^k+\sum_{k=1}^{\infty}kc_kx^k-\sum_{k=0}^{\infty}c_kx^k$$

$$=(-c_2-c_0)-6c_3x+\sum_{k=2}^{\infty}\left[-(k+2)(k+1)c_{k+2}+\left(k^2-1\right)c_k\right]x^k=0.$$

Thus

$$-2c_2-c_0=0$$

$$-6c_3=0$$

$$-(k+2)(k+1)c_{k+2}+(k-1)(k+1)c_k=0$$

and

$$c_2=-\frac{1}{2}c_0$$

$$c_3=0$$

$$c_{k+2}=\frac{k-1}{k+2}c_k,\quad k=2,3,4,\dots.$$

Choosing $c_0=1$ and $c_1=0$ we find

$$c_2=-\frac{1}{2}$$

$$c_3=c_5=c_7=\cdots=0$$

$$c_4=-\frac{1}{8}$$

and so on. For $c_0=0$ and $c_1=1$ we obtain

$$c_2=c_4=c_6=\cdots=0$$

$$c_3=c_5=c_7=\cdots=0.$$

Thus, two solutions are

$$y_1=1-\frac{1}{2}x^2-\frac{1}{8}x^4-\cdots\quad\text{and}\quad y_2=x.$$

15. Substituting $y=\sum_{n=0}^{\infty} c_n x^n$ into the differential equation we have

$$(x-1)y''-xy'+y=\underbrace{\sum_{n=2}^{\infty} n(n-1)c_n x^{n-1}}_{k=n-1}-\underbrace{\sum_{n=2}^{\infty} n(n-1)c_n x^{n-2}}_{k=n-2}-\underbrace{\sum_{n=1}^{\infty} nc_n x^n}_{k=n}+\underbrace{\sum_{n=0}^{\infty} c_n x^n}_{k=n}$$

$$=\sum_{k=1}^{\infty}(k+1)kc_{k+1}x^k-\sum_{k=0}^{\infty}(k+2)(k+1)c_{k+2}x^k-\sum_{k=1}^{\infty}kc_kx^k+\sum_{k=0}^{\infty}c_kx^k$$

$$=-2c_2+c_0+\sum_{k=1}^{\infty}[-(k+2)(k+1)c_{k+2}+(k+1)kc_{k+1}-(k-1)c_k]x^k=0.$$

Thus

$$-2c_2+c_0=0$$

$$-(k+2)(k+1)c_{k+2}+(k-1)kc_{k+1}-(k-1)c_k=0$$

and

$$c_2=\frac{1}{2}c_0$$

$$c_{k+2}=\frac{kc_{k+1}}{k+2}-\frac{(k-1)c_k}{(k+2)(k+1)},\qquad k=1,2,3,\ldots.$$

Choosing $c_0=1$ and $c_1=0$ we find

$$c_2=\frac{1}{2},\qquad c_3=\frac{1}{6},\qquad c_4=0$$

and so on. For $c_0=0$ and $c_1=1$ we obtain $c_2=c_3=c_4=\cdots=0$. Thus,

$$y=C_1\left(1+\frac{1}{2}x^2+\frac{1}{6}x^3+\cdots\right)+C_2x$$

and

$$y'=C_1\left(x+\frac{1}{2}x^2+\cdots\right)+C_2.$$

The initial conditions imply $C_1=-2$ and $C_2=6$, so

$$y=-2\left(1+\frac{1}{2}x^2+\frac{1}{6}x^3+\cdots\right)+6x=8x-2e^x.$$

18. Substituting $y = \sum_{n=0}^{\infty} c_n x^n$ into the differential equation we have

$$(x^2+1)y'' + 2xy' = \underbrace{\sum_{n=2}^{\infty} n(n-1)c_n x^n}_{k=n} + \underbrace{\sum_{n=2}^{\infty} n(n-1)c_n x^{n-2}}_{k=n-2} + \underbrace{\sum_{n=1}^{\infty} 2nc_n x^n}_{k=n}$$

$$= \sum_{k=2}^{\infty} k(k-1)c_k x^k + \sum_{k=0}^{\infty} (k+2)(k+1)c_{k+2} x^k + \sum_{k=1}^{\infty} 2kc_k x^k$$

$$= 2c_2 + (6c_3 + 2c_1)x + \sum_{k=2}^{\infty} [k(k+1)c_k + (k+2)(k+1)c_{k+2}]x^k = 0.$$

Thus

$$2c_2 = 0$$

$$6c_3 + 2c_1 = 0$$

$$k(k+1)c_k + (k+2)(k+1)c_{k+2} = 0$$

and

$$c_2 = 0$$

$$c_3 = -\frac{1}{3}c_1$$

$$c_{k+2} = -\frac{k}{k+2}c_k, \quad k = 2, 3, 4, \ldots .$$

Choosing $c_0 = 1$ and $c_1 = 0$ we find $c_3 = c_4 = c_5 = \cdots = 0$. For $c_0 = 0$ and $c_1 = 1$ we obtain

$$c_3 = -\frac{1}{3}$$

$$c_4 = c_6 = c_8 = \cdots = 0$$

$$c_5 = -\frac{1}{5}$$

$$c_7 = \frac{1}{7}$$

and so on. Thus

$$y = c_0 + c_1\left(x - \frac{1}{3}x^3 + \frac{1}{5}x^5 - \frac{1}{7}x^7 + \cdots\right)$$

and

$$y' = c_1\left(1 - x^2 + x^4 - x^6 + \cdots\right).$$

The initial conditions imply $c_0 = 0$ and $c_1 = 1$, so

$$y = x - \frac{1}{3}x^3 + \frac{1}{5}x^5 - \frac{1}{7}x^7 + \cdots .$$

21. Substituting $y = \sum_{n=0}^{\infty} c_n x^n$ into the differential equation we have

$$y'' + e^{-x}y = \sum_{n=2}^{\infty} n(n-1)c_n x^{n-2}$$

$$+ \left(1 - x + \frac{1}{2}x^2 - \frac{1}{6}x^3 + \frac{1}{24}x^4 - \cdots\right)\left(c_0 + c_1x + c_2x^2 + c_3x^3 + \cdots\right)$$

$$= \left[2c_2 + 6c_3x + 12c_4x^2 + 20c_5x^3 + \cdots\right] + \left[c_0 + (c_1 - c_0)x + \left(c_2 - c_1 + \frac{1}{2}c_0\right)x^2 + \cdots\right]$$

$$= (2c_2 + c_0) + (6c_3 + c_1 - c_0)x + \left(12c_4 + c_2 - c_1 + \frac{1}{2}c_0\right)x^2 + \cdots = 0.$$

Thus

$$2c_2 + c_0 = 0$$
$$6c_3 + c_1 - c_0 = 0$$
$$12c_4 + c_2 - c_1 + \frac{1}{2}c_0 = 0$$

and

$$c_2 = -\frac{1}{2}c_0$$
$$c_3 = -\frac{1}{6}c_1 + \frac{1}{6}c_0$$
$$c_4 = -\frac{1}{12}c_2 + \frac{1}{12}c_1 - \frac{1}{24}c_0.$$

Choosing $c_0 = 1$ and $c_1 = 0$ we find

$$c_2 = -\frac{1}{2}, \qquad c_3 = \frac{1}{6}, \qquad c_4 = 0$$

and so on. For $c_0 = 0$ and $c_1 = 1$ we obtain

$$c_2 = 0, \qquad c_3 = -\frac{1}{6}, \qquad c_4 = \frac{1}{12}.$$

Thus, two solutions are

$$y_1 = 1 - \frac{1}{2}x^2 + \frac{1}{6}x^3 + \cdots \quad \text{and} \quad y_2 = x - \frac{1}{6}x^3 + \frac{1}{12}x^4 + \cdots.$$

24. Substituting $y = \sum_{n=0}^{\infty} c_n x^n$ into the differential equation leads to

$$y'' - 4xy' - 4y = \underbrace{\sum_{n=2}^{\infty} n(n-1)c_n x^{n-2}}_{k=n-2} - \underbrace{\sum_{n=1}^{\infty} 4nc_n x^n}_{k=n} - \underbrace{\sum_{n=0}^{\infty} 4c_n x^n}_{k=n}$$

$$= \sum_{k=0}^{\infty}(k+2)(k+1)c_{k+2}x^k - \sum_{k=1}^{\infty} 4kc_k x^k - \sum_{k=0}^{\infty} 4c_k x^k$$

$$= 2c_2 - 4c_0 + \sum_{k=1}^{\infty}[(k+2)(k+1)c_{k+2} - 4(k+1)c_k]x^k$$

$$= e^x = 1 + \sum_{k=1}^{\infty}\frac{1}{k!}x^k.$$

Thus
$$2c_2 - 4c_0 = 1$$
$$(k+2)(k+1)c_{k+2} - 4(k+1)c_k = \frac{1}{k!}$$

and
$$c_2 = \frac{1}{2} + 2c_0$$
$$c_{k+2} = \frac{1}{(k+2)!} + \frac{4}{k+2}c_k, \qquad k = 1, 2, 3, \ldots.$$

Let c_0 and c_1 be arbitrary and iterate to find

$$c_2 = \frac{1}{2} + 2c_0$$

$$c_3 = \frac{1}{3!} + \frac{4}{3}c_1 = \frac{1}{3!} + \frac{4}{3}c_1$$

$$c_4 = \frac{1}{4!} + \frac{4}{4}c_2 = \frac{1}{4!} + \frac{1}{2} + 2c_0 = \frac{13}{4!} + 2c_0$$

$$c_5 = \frac{1}{5!} + \frac{4}{5}c_3 = \frac{1}{5!} + \frac{4}{5\cdot 3!} + \frac{16}{15}c_1 = \frac{17}{5!} + \frac{16}{15}c_1$$

$$c_6 = \frac{1}{6!} + \frac{4}{6}c_4 = \frac{1}{6!} + \frac{4\cdot 13}{6\cdot 4!} + \frac{8}{6}c_0 = \frac{261}{6!} + \frac{4}{3}c_0$$

$$c_7 = \frac{1}{7!} + \frac{4}{7}c_5 = \frac{1}{7!} + \frac{4\cdot 17}{7\cdot 5!} + \frac{64}{105}c_1 = \frac{409}{7!} + \frac{64}{105}c_1$$

and so on. The solution is

$$y = c_0 + c_1x + \left(\frac{1}{2} + 2c_0\right)x^2 + \left(\frac{1}{3!} + \frac{4}{3}c_1\right)x^3 - \left(\frac{13}{4!} + 2c_0\right)x^4 + \left(\frac{17}{5!} + \frac{16}{15}c_1\right)x^5$$

$$+ \left(\frac{261}{6!} + \frac{4}{3}c_0\right)x^6 + \left(\frac{409}{7!} + \frac{64}{105}c_1\right)x^7 + \cdots$$

$$= c_0\left[1 + 2x^2 + 2x^4 + \frac{4}{3}x^6 + \cdots\right] + c_1\left[x + \frac{4}{3}x^3 + \frac{16}{15}x^5 + \frac{64}{105}x^7 + \cdots\right]$$

$$+ \frac{1}{2}x^2 + \frac{1}{3!}x^3 + \frac{13}{4!}x^4 + \frac{17}{5!}x^5 + \frac{261}{6!}x^6 + \frac{409}{7!}x^7 + \cdots.$$

Exercises 6.3

3. Irregular singular point: $x = 3$; regular singular point: $x = -3$

6. Irregular singular point: $x = 5$; regular singular point: $x = 0$

9. Irregular singular point: $x = 0$; regular singular points: $x = 2, \pm 5$

12. Substituting $y = \sum_{n=0}^{\infty} c_n x^{n+r}$ into the differential equation and collecting terms, we obtain

$$\begin{aligned} 2xy'' + 5y' + xy &= \left(2r^2 + 3r\right) c_0 x^{r-1} + \left(2r^2 + 7r + 5\right) c_1 x^r \\ &\quad + \sum_{k=2}^{\infty} [2(k+r)(k+r-1)c_k + 5(k+r)c_k + c_{k-2}]x^{k+r-1} \\ &= 0, \end{aligned}$$

which implies

$$2r^2 + 3r = r(2r+3) = 0,$$

$$\left(2r^2 + 7r + 5\right) c_1 = 0,$$

and

$$(k+r)(2k+2r+3)c_k + c_{k-2} = 0.$$

The indicial roots are $r = -3/2$ and $r = 0$, so $c_1 = 0$. For $r = -3/2$ the recurrence relation is

$$c_k = -\frac{c_{k-2}}{(2k-3)k}, \quad k = 2, 3, 4, \ldots,$$

and

$$c_2 = -\frac{1}{2}c_0, \qquad c_3 = 0, \qquad c_4 = \frac{1}{40}c_0.$$

For $r = 0$ the recurrence relation is

$$c_k = -\frac{c_{k-2}}{k(2k+3)}, \quad k = 2, 3, 4, \ldots,$$

and

$$c_2 = -\frac{1}{14}c_0, \qquad c_3 = 0, \qquad c_4 = \frac{1}{616}c_0.$$

The general solution on $(0, \infty)$ is

$$y = C_1 x^{-3/2}\left(1 - \frac{1}{2}x^2 + \frac{1}{40}x^4 + \cdots\right) + C_2\left(1 - \frac{1}{14}x^2 + \frac{1}{616}x^4 + \cdots\right).$$

15. Substituting $y = \sum_{n=0}^{\infty} c_n x^{n+r}$ into the differential equation and collecting terms, we obtain

$$\begin{aligned} 3xy'' + (2-x)y' - y &= \left(3r^2 - r\right) c_0 x^{r-1} \\ &\quad + \sum_{k=1}^{\infty} [3(k+r-1)(k+r)c_k + 2(k+r)c_k - (k+r)c_{k-1}]x^{k+r-1} \\ &= 0, \end{aligned}$$

which implies $$3r^2 - r = r(3r-1) = 0$$

and $$(k+r)(3k+3r-1)c_k - (k+r)c_{k-1} = 0.$$

The indicial roots are $r = 0$ and $r = 1/3$. For $r = 0$ the recurrence relation is

$$c_k = \frac{c_{k-1}}{(3k-1)}, \quad k = 1, 2, 3, \ldots,$$

and $$c_1 = \frac{1}{2}c_0, \qquad c_2 = \frac{1}{10}c_0, \qquad c_3 = \frac{1}{80}c_0.$$

For $r = 1/3$ the recurrence relation is

$$c_k = \frac{c_{k-1}}{3k}, \quad k = 1, 2, 3, \ldots,$$

and $$c_1 = \frac{1}{3}c_0, \qquad c_2 = \frac{1}{18}c_0, \qquad c_3 = \frac{1}{162}c_0.$$

The general solution on $(0, \infty)$ is

$$y = C_1\left(1 + \frac{1}{2}x + \frac{1}{10}x^2 + \frac{1}{80}x^3 + \cdots\right) + C_2 x^{1/3}\left(1 + \frac{1}{3}x + \frac{1}{18}x^2 + \frac{1}{162}x^3 + \cdots\right).$$

18. Substituting $y = \sum_{n=0}^{\infty} c_n x^{n+r}$ into the differential equation and collecting terms, we obtain

$$x^2y'' + xy' + \left(x^2 - \frac{4}{9}\right)y = \left(r^2 - \frac{4}{9}\right)c_0x^r + \left(r^2 + 2r + \frac{5}{9}\right)c_1x^{r+1}$$
$$+ \sum_{k=2}^{\infty}\left[(k+r)(k+r-1)c_k + (k+r)c_k - \frac{4}{9}c_k + c_{k-2}\right]x^{k+r}$$
$$= 0,$$

which implies

$$r^2 - \frac{4}{9} = \left(r + \frac{2}{3}\right)\left(r - \frac{2}{3}\right) = 0,$$

$$\left(r^2 + 2r + \frac{5}{9}\right)c_1 = 0,$$

and $$\left[(k+r)^2 - \frac{4}{9}\right]c_k + c_{k-2} = 0.$$

The indicial roots are $r = -2/3$ and $r = 2/3$, so $c_1 = 0$. For $r = -2/3$ the recurrence relation is

$$c_k = -\frac{9c_{k-2}}{3k(3k-4)}, \quad k = 2, 3, 4, \ldots,$$

and $$c_2 = -\frac{3}{4}c_0, \qquad c_3 = 0, \qquad c_4 = \frac{9}{128}c_0.$$

For $r = 2/3$ the recurrence relation is

$$c_k = -\frac{9c_{k-2}}{3k(3k+4)}, \qquad k = 2, 3, 4, \ldots,$$

and
$$c_2 = -\frac{3}{20}c_0, \qquad c_3 = 0, \qquad c_4 = \frac{9}{1{,}280}c_0.$$

The general solution on $(0, \infty)$ is

$$y = C_1 x^{-2/3}\left(1 - \frac{3}{4}x^2 + \frac{9}{128}x^4 + \cdots\right) + C_2 x^{2/3}\left(1 - \frac{3}{20}x^2 + \frac{9}{1{,}280}x^4 + \cdots\right).$$

21. Substituting $y = \sum_{n=0}^{\infty} c_n x^{n+r}$ into the differential equation and collecting terms, we obtain

$$2x^2y'' - x(x-1)y' - y = \left(2r^2 - r - 1\right)c_0x^r$$
$$+ \sum_{k=1}^{\infty}[2(k+r)(k+r-1)c_k + (k+r)c_k - c_k - (k+r-1)c_{k-1}]x^{k+r}$$
$$= 0,$$

which implies
$$2r^2 - r - 1 = (2r+1)(r-1) = 0$$

and
$$[(k+r)(2k+2r-1) - 1]c_k - (k+r-1)2c_{k-1} = 0.$$

The indicial roots are $r = -1/2$ and $r = 1$. For $r = -1/2$ the recurrence relation is

$$c_k = \frac{c_{k-1}}{2k}, \qquad k = 1, 2, 3, \ldots,$$

and
$$c_1 = \frac{1}{2}c_0, \qquad c_2 = \frac{1}{8}c_0, \qquad c_3 = \frac{1}{48}c_0.$$

For $r = 1$ the recurrence relation is

$$c_k = \frac{c_{k-1}}{2k+3}, \qquad k = 1, 2, 3, \ldots,$$

and
$$c_1 = \frac{1}{5}c_0, \qquad c_2 = \frac{1}{35}c_0, \qquad c_3 = \frac{1}{315}c_0.$$

The general solution on $(0, \infty)$ is

$$y = C_1 x^{-1/2}\left(1 + \frac{1}{2}x + \frac{1}{8}x^2 + \frac{1}{48}x^3 + \cdots\right) + C_2 x\left(1 + \frac{1}{5}x + \frac{1}{35}x^2 + \frac{1}{315}x^3 + \cdots\right).$$

24. Substituting $y = \sum_{n=0}^{\infty} c_n x^{n+r}$ into the differential equation and collecting terms, we obtain

$$x^2y'' + xy' + \left(x^2 - \frac{1}{4}\right)y = \left(r^2 - \frac{1}{4}\right)c_0x^r + \left(r^2 + 2r + \frac{3}{4}\right)c_1x^{r+1}$$
$$+ \sum_{k=2}^{\infty}\left[(k+r)(k+r-1)c_k + (k+r)c_k - \frac{1}{4}c_k + c_{k-2}\right]x^{k+r}$$
$$= 0,$$

which implies

$$r^2 - \frac{1}{4} = \left(r - \frac{1}{2}\right)\left(r + \frac{1}{2}\right) = 0,$$

$$\left(r^2 + 2r + \frac{3}{4}\right) c_1 = 0,$$

and
$$\left[(k+r)^2 - \frac{1}{4}\right] c_k + c_{k-2} = 0.$$

The indicial roots are $r_1 = 1/2$ and $r_2 = -1/2$, so $c_1 = 0$. For $r_1 = 1/2$ the recurrence relation is

$$c_k = -\frac{c_{k-2}}{k(k+1)}, \quad k = 2, 3, 4, \ldots,$$

and
$$c_2 = -\frac{1}{3!}c_0$$

$$c_3 = c_5 = c_7 = \cdots = 0$$

$$c_4 = \frac{1}{5!}c_0$$

$$c_{2n} = \frac{(-1)^n}{(2n+1)!}c_0.$$

For $r_2 = -1/2$ the recurrence relation is

$$c_k = -\frac{c_{k-2}}{k(k-1)}, \quad k = 2, 3, 4, \ldots,$$

and
$$c_2 = -\frac{1}{2!}c_0$$

$$c_3 = c_5 = c_7 = \cdots = 0$$

$$c_4 = \frac{1}{4!}c_0$$

$$c_{2n} = \frac{(-1)^n}{(2n)!}c_0.$$

The general solution on $(0, \infty)$ is

$$y = C_1 x^{1/2} \sum_{n=0}^{\infty} \frac{(-1)^n}{(2n+1)!} x^{2n} + C_2 x^{-1/2} \sum_{n=0}^{\infty} \frac{(-1)^n}{(2n)!} x^{2n}$$

$$= C_1 x^{-1/2} \sum_{n=0}^{\infty} \frac{(-1)^n}{(2n+1)!} x^{2n+1} + C_2 x^{-1/2} \sum_{n=0}^{\infty} \frac{(-1)^n}{(2n)!} x^{2n}$$

$$= x^{-1/2}[C_1 \sin x + C_2 \cos x].$$

27. Substituting $y = \sum_{n=0}^{\infty} c_n x^{n+r}$ into the differential equation and collecting terms, we obtain

$$xy'' + (1-x)y' - y = r^2 c_0 x^{r-1} + \sum_{k=0}^{\infty} [(k+r)(k+r-1)c_k + (k+r)c_k - (k+r)c_{k-1}]x^{k+r-1} = 0,$$

which implies $r^2 = 0$ and

$$(k+r)^2 c_k - (k+r)c_{k-1} = 0.$$

The indicial roots are $r_1 = r_2 = 0$ and the recurrence relation is

$$c_k = \frac{c_{k-1}}{k}, \quad k = 1, 2, 3, \ldots .$$

One solution is

$$y_1 = c_0\left(1 + x + \frac{1}{2}x^2 + \frac{1}{3!}x^3 + \cdots\right) = c_0 e^x.$$

A second solution is

$$y_2 = y_1 \int \frac{e^{-\int (1/x - 1)dx}}{e^{2x}}\,dx = e^x \int \frac{e^x/x}{e^{2x}}\,dx = e^x \int \frac{1}{x}e^{-x}dx$$

$$= e^x \int \frac{1}{x}\left(1 - x + \frac{1}{2}x^2 - \frac{1}{3!}x^3 + \cdots\right)dx = e^x \int \left(\frac{1}{x} - 1 + \frac{1}{2}x - \frac{1}{3!}x^2 + \cdots\right)dx$$

$$= e^x\left[\ln x - x + \frac{1}{2\cdot 2}x^2 - \frac{1}{3\cdot 3!}x^3 + \cdots\right] = e^x \ln x - e^x \sum_{n=1}^{\infty} \frac{(-1)^{n+1}}{n\cdot n!}x^n.$$

The general solution on $(0, \infty)$ is

$$y = C_1 e^x + C_2 e^x\left(\ln x - \sum_{n=1}^{\infty} \frac{(-1)^{n+1}}{n\cdot n!}x^n\right).$$

30. Substituting $y = \sum_{n=0}^{\infty} c_n x^{n+r}$ into the differential equation and collecting terms, we obtain

$$xy'' - xy' + y = \left(r^2 - r\right)c_0 x^{r-1} + \sum_{k=0}^{\infty} [(k+r+1)(k+r)c_{k+1} - (k+r)c_k + c_k]x^{k+r} = 0$$

which implies

$$r^2 - r = r(r-1) = 0$$

and

$$(k+r+1)(k+r)c_{k+1} - (k+r-1)c_k = 0.$$

The indicial roots are $r_1 = 1$ and $r_2 = 0$. For $r_1 = 1$ the recurrence relation is

$$c_{k+1} = \frac{kc_k}{(k+2)(k+1)}, \quad k = 0, 1, 2, \ldots,$$

and one solution is $y_1 = c_0 x$. A second solution is

$$y_2 = x\int \frac{e^{-\int -dx}}{x^2}\,dx = x\int \frac{e^x}{x^2}\,dx = x\int \frac{1}{x^2}\left(1 + x + \frac{1}{2}x^2 + \frac{1}{3!}x^3 + \cdots\right)dx$$

$$= x\int\left(\frac{1}{x^2} + \frac{1}{x} + \frac{1}{2} + \frac{1}{3!}x + \frac{1}{4!}x^2 + \cdots\right)dx = x\left[-\frac{1}{x} + \ln x + \frac{1}{2}x + \frac{1}{12}x^2 + \frac{1}{72}x^3 + \cdots\right]$$

$$= x\ln x - 1 + \frac{1}{2}x^2 + \frac{1}{12}x^3 + \frac{1}{72}x^4 + \cdots.$$

The general solution on $(0, \infty)$ is

$$y = C_1 x + C_2 y_2(x).$$

Exercises 6.4

3. Since $\nu^2 = 25/4$ the general solution is $y = c_1 J_{5/2}(x) + c_2 J_{-5/2}(x)$.

6. Since $\nu^2 = 4$ the general solution is $y = c_1 J_2(x) + c_2 Y_2(x)$.

9. If $y = x^{-1/2}v(x)$ then

$$y' = x^{-1/2}v'(x) - \frac{1}{2}x^{-3/2}v(x),$$

$$y'' = x^{-1/2}v''(x) - x^{-3/2}v'(x) + \frac{3}{4}x^{-5/2}v(x),$$

and

$$x^2y'' + 2xy' + \lambda^2x^2y = x^{3/2}v'' + x^{1/2}v' + \left(\lambda^2x^{3/2} - \frac{1}{4}x^{-1/2}\right)v.$$

Multiplying by $x^{1/2}$ we obtain

$$x^2v'' + xv' + \left(\lambda^2x^2 - \frac{1}{4}\right)v = 0,$$

whose solution is $v = c_1 J_{1/2}(\lambda x) + c_2 J_{-1/2}(\lambda x)$. Then $y = c_1 x^{-1/2} J_{1/2}(\lambda x) + c_2 x^{-1/2} J_{-1/2}(\lambda x)$.

12. From $y = \sqrt{x}\, J_\nu(\lambda x)$ we find

$$y' = \lambda\sqrt{x}\, J_\nu'(\lambda x) + \frac{1}{2}x^{-1/2}J_\nu(\lambda x)$$

and

$$y'' = \lambda^2\sqrt{x}\, J_\nu''(\lambda x) + \lambda x^{-1/2}J_\nu'(\lambda x) - \frac{1}{4}x^{-3/2}J_\nu(\lambda x).$$

Substituting into the differential equation, we have

$$x^2y'' + \left(\lambda^2x^2 - \nu^2 + \frac{1}{4}\right)y = \sqrt{x}\left[\lambda^2x^2J''_\nu(\lambda x) + \lambda xJ'_\nu(\lambda x) + \left(\lambda^2x^2 - \nu^2\right)J_\nu(\lambda x)\right]$$

$$= \sqrt{x}\cdot 0 \qquad \text{(since } J_n \text{ is a solution of Bessel's equation)}$$

$$= 0.$$

Therefore, $\sqrt{x}\,J_\nu(\lambda x)$ is a solution of the original equation.

15. From Problem 10 with $n = -1$ we find $y = x^{-1}J_{-1}(x)$. From Problem 11 with $n = 1$ we find $y = x^{-1}J_1(x) = -x^{-1}J_{-1}(x)$.

18. From Problem 10 with $n = 3$ we find $y = x^3J_3(x)$. From Problem 11 with $n = -3$ we find $y = x^3J_{-3}(x) = -x^3J_3(x)$.

21. The recurrence relation follows from

$$xJ_{\nu+1}(x) + xJ_{\nu-1}(x) = \sum_{n=0}^{\infty}\frac{(-1)^{n-1}2n}{n!\Gamma(1+\nu+n)}\left(\frac{x}{2}\right)^{2n+\nu} + \sum_{n=0}^{\infty}\frac{(-1)^n2(\nu+n)}{n!\Gamma(1+\nu+n)}\left(\frac{x}{2}\right)^{2n+\nu}$$

$$= \sum_{n=0}^{\infty}\frac{(-1)^n2\nu}{n!\Gamma(1+\nu+n)}\left(\frac{x}{2}\right)^{2n+\nu} = 2\nu J_\nu(x).$$

24. From (14) in the text we obtain $J_0'(x) = -J_1(x)$ and from (15) in the text we obtain $J_0'(x) = J_{-1}(x)$. Thus

$$J_0'(x) = J_{-1}(x) = -J_1(x).$$

27. Since

$$\Gamma\left(1 - \frac{1}{2} + n\right) = \frac{(2n-1)!}{(n-1)!2^{2n-1}}$$

we obtain

$$J_{-1/2}(x) = \sum_{n=0}^{\infty}\frac{(-1)^n2^{1/2}x^{-1/2}}{2n(2n-1)!\sqrt{\pi}}x^{2n} = \sqrt{\frac{2}{\pi x}}\cos x.$$

30. By Problem 21 we obtain $3J_{3/2}(x) = xJ_{5/2}(x) + xJ_{1/2}(x)$ so that

$$J_{5/2}(x) = \sqrt{\frac{2}{\pi x}}\left(\frac{3\sin x}{x^2} - \frac{3\cos x}{x} - \sin x\right).$$

33. By Problem 21 we obtain $-5J_{-5/2}(x) = xJ_{-3/2}(x) + xJ_{-7/2}(x)$ so that

$$J_{-7/2}(x) = \sqrt{\frac{2}{\pi x}}\left(\frac{-15\cos x}{x^3} - \frac{15\sin x}{x^2} + \frac{6\cos x}{x} + \sin x\right).$$

36. If $y_1 = J_0(x)$ then using the formula for the second solution of a linear homogeneous second-order differential equation gives

$$\begin{aligned}
y_2 &= J_0(x)\int \frac{e^{-\int dx/x}}{(J_0(x))^2}\,dx \\
&= J_0(x)\int \frac{dx}{x\left(1-\frac{x^2}{4}+\frac{x^4}{64}-\frac{x^6}{2304}+\cdots\right)^2}\,dx \\
&= J_0(x)\int\left(\frac{1}{x}+\frac{x}{2}+\frac{5x^3}{32}+\frac{23x^5}{576}+\cdots\right)dx \\
&= J_0(x)\left(\ln x+\frac{x^2}{4}+\frac{5x^4}{128}+\frac{23x^6}{3456}+\cdots\right) \\
&= J_0(x)\ln x+\left(1-\frac{x^2}{4}+\frac{x^4}{64}-\frac{x^6}{2304}+\cdots\right)\left(\frac{x^2}{4}+\frac{5x^4}{128}+\frac{23x^6}{3456}+\cdots\right) \\
&= J_0(x)\ln x+\frac{x^2}{4}-\frac{3x^4}{128}+\frac{11x^6}{13824}-\cdots.
\end{aligned}$$

39. Letting

$$s=\frac{2}{\alpha}\sqrt{\frac{k}{m}}\,e^{-\alpha t/2},$$

we have

$$\begin{aligned}
\frac{dx}{dt} &= \frac{dx}{ds}\frac{ds}{dt} = \frac{dx}{dt}\left[\frac{2}{\alpha}\sqrt{\frac{k}{m}}\left(-\frac{\alpha}{2}\right)e^{-\alpha t/2}\right] \\
&= \frac{dx}{ds}\left(-\sqrt{\frac{k}{m}}\,e^{-\alpha t/2}\right)
\end{aligned}$$

and

$$\begin{aligned}
\frac{d^2x}{dt^2} &= \frac{d}{dt}\left(\frac{dx}{dt}\right) = \frac{dx}{ds}\left(\frac{\alpha}{2}\sqrt{\frac{k}{m}}\,e^{-\alpha t/2}\right)+\frac{d}{dt}\left(\frac{dx}{ds}\right)\left(-\sqrt{\frac{k}{m}}\,e^{-\alpha t/2}\right) \\
&= \frac{dx}{ds}\left(\frac{\alpha}{2}\sqrt{\frac{k}{m}}\,e^{-\alpha t/2}\right)+\frac{d^2x}{ds^2}\frac{ds}{dt}\left(-\sqrt{\frac{k}{m}}\,e^{-\alpha t/2}\right) \\
&= \frac{dx}{ds}\left(\frac{\alpha}{2}\sqrt{\frac{k}{m}}\,e^{-\alpha t/2}\right)+\frac{d^2x}{ds^2}\left(\frac{k}{m}\,e^{-\alpha t}\right).
\end{aligned}$$

Then

$$m\frac{d^2x}{dt^2}+ke^{-\alpha t}x = ke^{-\alpha t}\frac{d^2x}{ds^2}+\frac{m\alpha}{2}\sqrt{\frac{k}{m}}\,e^{-\alpha t/2}\frac{dx}{dt}+ke^{-\alpha t}x=0.$$

Multiplying by $2^2/\alpha^2 m$ we have

$$\frac{2^2}{\alpha^2}\frac{k}{m}e^{-\alpha t}\frac{d^2x}{ds^2}+\frac{2}{\alpha}\sqrt{\frac{k}{m}}\,e^{-\alpha t/2}\frac{dx}{dt}+\frac{2}{\alpha^2}\frac{k}{m}e^{-\alpha t}x=0$$

or, since $s=(2/\alpha)\sqrt{k/m}\,e^{-\alpha t/2}$,

$$s^2\frac{d^2x}{ds^2}+s\frac{dx}{ds}+s^2x=0.$$

42. The general solution of Bessel's equation is

$$w(t)=c_1J_{1/3}(t)+c_2J_{-1/3}(t), \qquad t>0.$$

Thus, the general solution of Airy's equation for $x>0$ is

$$y=x^{1/2}w\left(\frac{2}{3}\alpha x^{3/2}\right)=c_1x^{1/2}J_{1/3}\left(\frac{2}{3}\alpha x^{3/2}\right)+c_2x^{1/2}J_{-1/3}\left(\frac{2}{3}\alpha x^{3/2}\right).$$

45. (a) Using the expressions for the two linearly independent power series solutions, $y_1(x)$ and $y_2(x)$, given in the text we obtain

$$P_6(x)=\frac{1}{16}\left(231x^6-315x^4+105x^2-5\right)$$

and

$$P_7(x)=\frac{1}{16}\left(429x^7-693x^5+315x^3-35x\right).$$

(b) $P_6(x)$ satisfies $\left(1-x^2\right)y''-2xy'+42y=0$ and $P_7(x)$ satisfies $\left(1-x^2\right)y''-2xy'+56y=0$.

48. The polynomials are shown in (19) on page 251 in the text.

51. The recurrence relation can be written

$$P_{k+1}(x)=\frac{2k+1}{k+1}xP_k(x)-\frac{k}{k+1}P_{k-1}(x), \qquad k=2,\ 3,\ 4,\ \ldots\,.$$

$$k=1:\ \ P_2(x)=\frac{3}{2}x^2-\frac{1}{2}$$

$$k=2:\ \ P_3(x)=\frac{5}{3}x\left(\frac{3}{2}x^2-\frac{1}{2}\right)-\frac{2}{3}x=\frac{5}{2}x^3-\frac{3}{2}x$$

$$k=3:\ \ P_4(x)=\frac{7}{4}x\left(\frac{5}{2}x^3-\frac{3}{2}x\right)-\frac{3}{4}\left(\frac{3}{2}x^2-\frac{1}{2}\right)=\frac{35}{8}x^4-\frac{30}{8}x^2+\frac{3}{8}$$

$$k=4:\ \ P_5(x)=\frac{9}{5}x\left(\frac{35}{8}x^4-\frac{30}{8}x^2+\frac{3}{8}\right)-\frac{4}{5}\left(\frac{5}{2}x^3-\frac{3}{2}x\right)=\frac{63}{8}x^5-\frac{35}{4}x^3+\frac{15}{8}x$$

$$k=5:\ \ P_6(x)=\frac{11}{6}x\left(\frac{63}{8}x^5-\frac{35}{4}x^3+\frac{15}{8}x\right)-\frac{5}{6}\left(\frac{35}{8}x^4-\frac{30}{8}x^2+\frac{3}{8}\right)=\frac{231}{16}x^6-\frac{315}{16}x^4+\frac{105}{16}x-\frac{5}{16}$$

Chapter 6 Review Exercises

3. Since

$$P(x) = \frac{1}{x(x-5)^2} \quad \text{and} \quad Q(x) = 0$$

the regular singular point is $x = 0$ and the irregular singular point is $x = 5$.

6. Since

$$P(x) = \frac{1}{x\left(x^2+1\right)^3} \quad \text{and} \quad Q(x) = -\frac{8}{\left(x^2+1\right)^3}$$

the regular singular point is $x = 0$. The irregular singular points are $x = i$ and $x = -i$.

9. Substituting $y = \sum_{n=0}^{\infty} c_n x^n$ into the differential equation we have

$$y'' - xy' - y = \underbrace{\sum_{n=2}^{\infty} n(n-1)c_n x^{n-2}}_{k=n-2} - \underbrace{\sum_{n=1}^{\infty} nc_n x^n}_{k=n} - \underbrace{\sum_{n=0}^{\infty} c_n x^n}_{k=n}$$

$$= \sum_{k=0}^{\infty} (k+2)(k+1)c_{k+2}x^k - \sum_{k=1}^{\infty} kc_k x^k - \sum_{k=0}^{\infty} c_k x^k$$

$$= 2c_2 - c_0 + \sum_{k=1}^{\infty} [(k+2)(k+1)c_{k+2} - (k+1)c_k]x^k = 0.$$

Thus

$$2c_2 - c_0 = 0$$

$$(k+2)(k+1)c_{k+2} - (k+1)c_k = 0$$

and

$$c_2 = \frac{1}{2}c_0$$

$$c_{k+2} = \frac{1}{k+2}c_k, \quad k = 1, 2, 3, \ldots.$$

Choosing $c_0 = 1$ and $c_1 = 0$ we find

$$c_2 = \frac{1}{2}$$

$$c_3 = c_5 = c_7 = \cdots = 0$$

$$c_4 = \frac{1}{8}$$

$$c_6 = \frac{1}{48}$$

and so on. For $c_0 = 0$ and $c_1 = 1$ we obtain

$$c_2 = c_4 = c_6 = \cdots = 0$$
$$c_3 = \frac{1}{3}$$
$$c_5 = \frac{1}{15}$$
$$c_7 = \frac{1}{105}$$

and so on. Thus, two solutions are

$$y_1 = 1 + \frac{1}{2}x^2 + \frac{1}{8}x^4 + \frac{1}{48}x^6 + \cdots$$

and

$$y_2 = x + \frac{1}{3}x^3 + \frac{1}{15}x^5 + \frac{1}{105}x^7 + \cdots.$$

12. Substituting $y = \sum_{n=0}^{\infty} c_n x^n$ into the differential equation we have

$$\begin{aligned}(\cos x)y'' + y &= \left(1 - \frac{1}{2}x^2 + \frac{1}{24}x^4 - \frac{1}{720}x^6 + \cdots\right)(2c_2 + 6c_3x + 12c_4x^2 + 20c_5x^3 + 30c_6x^4 + \cdots) \\ &\quad + \sum_{n=0}^{\infty} c_n x^n \\ &= \left[2c_2 + 6c_3x + (12c_4 - c_2)x^2 + (20c_5 - 3c_3)x^3 + \left(30c_6 - 6c_4 + \frac{1}{12}c_2\right)x^4 + \cdots\right] \\ &\quad + [c_0 + c_1x + c_2x^2 + c_3x^3 + c_4x^4 + \cdots] \\ &= (c_0 + 2c_2) + (c_1 + 6c_3)x + 12c_4x^2 + (20c_5 - 2c_3)x^3 + \left(30c_6 - 5c_4 + \frac{1}{12}c_2\right)x^4 + \cdots \\ &= 0.\end{aligned}$$

Thus

$$c_0 + 2c_2 = 0$$
$$c_1 + 6c_3 = 0$$
$$12c_4 = 0$$
$$20c_5 - 2c_3 = 0$$
$$30c_6 - 5c_4 + \frac{1}{12}c_2 = 0$$

and

$$c_2 = -\frac{1}{2}c_0$$
$$c_3 = -\frac{1}{6}c_1$$

$$c_4 = 0$$
$$c_5 = \frac{1}{10}c_3$$
$$c_6 = \frac{1}{6}c_4 - \frac{1}{360}c_2.$$

Choosing $c_0 = 1$ and $c_1 = 0$ we find

$$c_2 = -\frac{1}{2}, \quad c_3 = 0, \quad c_4 = 0, \quad c_5 = 0, \quad c_6 = \frac{1}{720}$$

and so on. For $c_0 = 0$ and $c_1 = 1$ we find

$$c_2 = 0, \quad c_3 = -\frac{1}{6}, \quad c_4 = 0, \quad c_5 = -\frac{1}{60}, \quad c_6 = 0$$

and so on. Thus, two solutions are

$$y_1 = 1 - \frac{1}{2}x^2 + \frac{1}{720}x^6 + \cdots \quad \text{and} \quad y_2 = x - \frac{1}{6}x^3 - \frac{1}{60}x^5 + \cdots.$$

15. Substituting $y = \sum_{n=0}^{\infty} c_n x^{n+r}$ into the differential equation we obtain

$$\begin{aligned} 2x^2y'' + xy' - (x+1)y \\ &= \left(2r^2 - r - 1\right)c_0x^r + \sum_{k=1}^{\infty}[2(k+r)(k+r-1)c_k + (k+r)c_k - c_k - c_{k-1}]x^{k+r} \\ &= 0 \end{aligned}$$

which implies $$2r^2 - r - 1 = (2r+1)(r-1) = 0$$

and $$[(k+r)(2k+2r-1) - 1]c_k - c_{k-1} = 0$$

The indicial roots are $r = 1$ and $r = -1/2$. For $r = 1$ the recurrence relation is

$$c_k = \frac{c_{k-1}}{k(2k+3)}, \quad k = 1, 2, 3, \ldots,$$

so $$c_1 = \frac{1}{5}c_0, \quad c_2 = \frac{1}{70}c_0, \quad c_3 = \frac{1}{1{,}890}c_0.$$

For $r = -1/2$ the recurrence relation is

$$c_k = \frac{c_{k-1}}{k(2k-3)}, \quad k = 1, 2, 3, \ldots,$$

so $$c_1 = -c_0, \quad c_2 = -\frac{1}{2}c_0, \quad c_3 = -\frac{1}{18}c_0.$$

Two linearly independent solutions are

$$y_1 = C_1 x\left(1 + \frac{1}{5}x + \frac{1}{70}x^2 + \frac{1}{1{,}890}x^3 + \cdots\right)$$

and

$$y_2 = C_2 x^{-1/2}\left(1 - x - \frac{1}{2}x^2 - \frac{1}{18}x^3 - \cdots\right).$$

18. Substituting $y = \sum_{n=0}^{\infty} c_n x^{n+r}$ into the differential equation we obtain

$$x^2y'' - xy' + \left(x^2+1\right)y = \left(r^2 - 2r + 1\right)c_0x^r + r^2c_1x^{r+1} + \sum_{k=2}^{\infty}[(k+r)(k+r-1)c_k - (k+r)c_k + c_k + c_{k-2}]x^{k+r} = 0$$

which implies

$$r^2 - 2r + 1 = (r-1)^2 = 0$$
$$r^2c_1 = 0$$
$$[(k+r)(k+r-2)+1]c_k + c_{k-2} = 0.$$

The indicial roots are $r_1 = r_2 = 1$, so $c_1 = 0$ and

$$c_k = -\frac{c_{k-2}}{k^2}, \quad k = 2, 3, 4, \ldots.$$

Thus

$$c_2 = -\frac{1}{4}c_0$$
$$c_3 = c_5 = c_7 = \cdots = 0$$
$$c_4 = \frac{1}{64}c_0$$
$$c_6 = -\frac{1}{2{,}304}c_0$$

and one solution is

$$y_1 = c_0 x\left(1 - \frac{1}{4}x^2 + \frac{1}{64}x^4 - \frac{1}{2{,}304}x^6 + \cdots\right).$$

A second solution is

$$y_2 = y_1\int \frac{e^{dx/x}}{y_1^2}\,dx = y_1\int \frac{x\,dx}{x^2\left(1 - \frac{1}{4}x^2 + \frac{1}{64}x^4 - \frac{1}{2304}x^6 + \cdots\right)^2}$$

$$= y_1\int \frac{dx}{x\left(1 - \frac{1}{2}x^2 + \frac{3}{32}x^4 - \frac{5}{576}x^6 + \cdots\right)}$$

$$= y_1\int \frac{1}{x}\left(1 + \frac{1}{2}x^2 + \frac{5}{32}x^4 + \frac{23}{576}x^6 + \cdots\right)dx$$

$$= y_1 \int \left(\frac{1}{x} + \frac{1}{2}x + \frac{5}{32}x^3 + \frac{23}{576}x^5 + \cdots \right) dx$$

$$= y_1 \ln x + y_1 \left(\frac{1}{4}x^2 + \frac{5}{128}x^4 + \frac{23}{3{,}456}x^6 + \cdots \right).$$

7 The Laplace Transform

Exercises 7.1

3. $\mathscr{L}\{f(t)\} = \int_0^1 te^{-st}dt + \int_1^\infty e^{-st}dt = \left(-\frac{1}{s}te^{-st} - \frac{1}{s^2}e^{-st}\right)\Big|_0^1 - \frac{1}{s}e^{-st}\Big|_1^\infty$

$= \left(-\frac{1}{s}e^{-s} - \frac{1}{s^2}e^{-s}\right) - \left(0 - \frac{1}{s^2}\right) - \frac{1}{s}(0 - e^{-s}) = \frac{1}{s^2}(1 - e^{-s}), \quad s > 0$

6. $\mathscr{L}\{f(t)\} = \int_{\pi/2}^\infty (\cos t)e^{-st}dt = \left(-\frac{s}{s^2+1}e^{-st}\cos t + \frac{1}{s^2+1}e^{-st}\sin t\right)\Big|_{\pi/2}^\infty$

$= 0 - \left(0 + \frac{1}{s^2+1}e^{-\pi s/2}\right) = -\frac{1}{s^2+1}e^{-\pi s/2}, \quad s > 0$

9. $f(t) = \begin{cases} 1-t, & 0 < t < 1 \\ 0, & t > 0 \end{cases}$

$\mathscr{L}\{f(t)\} = \int_0^1 (1-t)e^{-st}\,dt = \left(-\frac{1}{s}(1-t)e^{-st} + \frac{1}{s^2}\,e^{-st}\right)\Big|_0^1 = \frac{1}{s^2}\,e^{-s} + \frac{1}{s} - \frac{1}{s^2}, \quad s > 0$

12. $\mathscr{L}\{f(t)\} = \int_0^\infty e^{-2t-5}e^{-st}dt = e^{-5}\int_0^\infty e^{-(s+2)t}dt = -\frac{e^{-5}}{s+2}e^{-(s+2)t}\Big|_0^\infty = \frac{e^{-5}}{s+2}, \quad s > -2$

15. $\mathscr{L}\{f(t)\} = \int_0^\infty e^{-t}(\sin t)e^{-st}dt = \int_0^\infty (\sin t)e^{-(s+1)t}dt$

$= \left(\frac{-(s+1)}{(s+1)^2+1}e^{-(s+1)t}\sin t - \frac{1}{(s+1)^2+1}e^{-(s+1)t}\cos t\right)\Big|_0^\infty$

$= \frac{1}{(s+1)^2+1} = \frac{1}{s^2+2s+2}, \quad s > -1$

18. $\mathscr{L}\{f(t)\} = \int_0^\infty t(\sin t)e^{-st}dt$

$= \left[\left(-\frac{t}{s^2+1} - \frac{2s}{(s^2+1)^2}\right)(\cos t)e^{-st} - \left(\frac{st}{s^2+1} + \frac{s^2-1}{(s^2+1)^2}\right)(\sin t)e^{-st}\right]_0^\infty$

$= \frac{2s}{(s^2+1)^2}, \quad s > 0$

21. $\mathscr{L}\{4t - 10\} = \frac{4}{s^2} - \frac{10}{s}$

24. $\mathscr{L}\{-4t^2 + 16t + 9\} = -4\frac{2}{s^3} + \frac{16}{s^2} + \frac{9}{s}$

27. $\mathscr{L}\{1+e^{4t}\} = \dfrac{1}{s} + \dfrac{1}{s-4}$

30. $\mathscr{L}\{e^{2t} - 2 + e^{-2t}\} = \dfrac{1}{s-2} - \dfrac{2}{s} + \dfrac{1}{s+2}$

33. $\mathscr{L}\{\sinh kt\} = \dfrac{k}{s^2 - k^2}$

36. $\mathscr{L}\{e^{-t}\cosh t\} = \mathscr{L}\left\{e^{-t}\dfrac{e^t + e^{-t}}{2}\right\} = \mathscr{L}\left\{\dfrac{1}{2} + \dfrac{1}{2}e^{-2t}\right\} = \dfrac{1}{2s} + \dfrac{1}{2(s+2)}$

39. Let $u = st$ so that $du = s\,dt$ and $\mathscr{L}\{t^\alpha\} = \int_0^\infty e^{-st}t^\alpha dt = \int_0^\infty e^{-u}\left(\dfrac{u}{s}\right)^\alpha \dfrac{1}{s}\,du = \dfrac{1}{s^{\alpha+1}}\Gamma(\alpha+1)$ for $\alpha > -1$.

42. $\mathscr{L}\{t^{3/2}\} = \dfrac{\Gamma(5/2)}{s^{5/2}} = \dfrac{3\sqrt{\pi}}{4s^{5/2}}$

Exercises 7.2

3. $\mathscr{L}^{-1}\left\{\dfrac{1}{s^2} - \dfrac{48}{s^5}\right\} = \mathscr{L}^{-1}\left\{\dfrac{1}{s^2} - \dfrac{48}{24}\cdot\dfrac{4!}{s^5}\right\} = t - 2t^4$

6. $\mathscr{L}^{-1}\left\{\dfrac{(s+2)^2}{s^3}\right\} = \mathscr{L}^{-1}\left\{\dfrac{1}{s} + 4\cdot\dfrac{1}{s^2} + 2\cdot\dfrac{2}{s^3}\right\} = 1 + 4t + 2t^2$

9. $\mathscr{L}^{-1}\left\{\dfrac{1}{4s+1}\right\} = \mathscr{L}^{-1}\left\{\dfrac{1}{4}\cdot\dfrac{1}{s+1/4}\right\} = \dfrac{1}{4}e^{-t/4}$

12. $\mathscr{L}^{-1}\left\{\dfrac{10s}{s^2+16}\right\} = 10\cos 4t$

15. $\mathscr{L}^{-1}\left\{\dfrac{1}{s^2-16}\right\} = \mathscr{L}^{-1}\left\{\dfrac{1/8}{s-4} - \dfrac{1/8}{s+4}\right\} = \dfrac{1}{8}e^{4t} - \dfrac{1}{8}e^{-4t} = \dfrac{1}{4}\sinh 4t$

18. $\mathscr{L}^{-1}\left\{\dfrac{s+1}{s^2+2}\right\} = \mathscr{L}^{-1}\left\{\dfrac{s}{s^2+2} + \dfrac{1}{\sqrt{2}}\cdot\dfrac{\sqrt{2}}{s^2+2}\right\} = \cos\sqrt{2}t + \dfrac{1}{\sqrt{2}}\sin\sqrt{2}t$

21. $\mathscr{L}^{-1}\left\{\dfrac{s}{s^2+2s-3}\right\} = \mathscr{L}^{-1}\left\{\dfrac{1}{4}\cdot\dfrac{1}{s-1} + \dfrac{3}{4}\cdot\dfrac{1}{s+3}\right\} = \dfrac{1}{4}e^t + \dfrac{3}{4}e^{-3t}$

24. $\mathscr{L}^{-1}\left\{\dfrac{s-3}{(s-\sqrt{3})(s+\sqrt{3})}\right\} = \mathscr{L}^{-1}\left\{\dfrac{s}{s^2-3} - \sqrt{3}\cdot\dfrac{\sqrt{3}}{s^2-3}\right\} = \cosh\sqrt{3}t - \sqrt{3}\sinh\sqrt{3}t$

27. $\mathscr{L}^{-1}\left\{\dfrac{2s+4}{(s-2)(s^2+4s+3)}\right\} = \mathscr{L}^{-1}\left\{\dfrac{8}{15}\cdot\dfrac{1}{s-2} - \dfrac{1}{3}\cdot\dfrac{1}{s+1} - \dfrac{1}{5}\cdot\dfrac{1}{s+3}\right\} = \dfrac{8}{15}e^{2t} - \dfrac{1}{3}e^{-t} - \dfrac{1}{5}e^{-3t}$

30. $\mathscr{L}^{-1}\left\{\dfrac{s-1}{s^2(s^2+1)}\right\}=\mathscr{L}^{-1}\left\{\dfrac{1}{s}-\dfrac{1}{s^2}-\dfrac{s}{s^2+1}+\dfrac{1}{s^2+1}\right\}=1-t-\cos t+\sin t$

33. $\mathscr{L}^{-1}\left\{\dfrac{1}{(s^2+1)(s^2+4)}\right\}=\mathscr{L}^{-1}\left\{\dfrac{1}{3}\cdot\dfrac{1}{s^2+1}-\dfrac{1}{6}\cdot\dfrac{2}{s^2+4}\right\}=\dfrac{1}{3}\sin t-\dfrac{1}{6}\sin 2t$

Exercises 7.3

3. $\mathscr{L}\left\{t^3e^{-2t}\right\}=\dfrac{3!}{(s+2)^4}$

6. $\mathscr{L}\left\{e^{-2t}\cos 4t\right\}=\dfrac{s+2}{(s+2)^2+16}$

9. $\mathscr{L}\left\{t\left(e^t+e^{2t}\right)^2\right\}=\mathscr{L}\left\{te^{2t}+2te^{3t}+te^{4t}\right\}=\dfrac{1}{(s-2)^2}+\dfrac{2}{(s-3)^2}+\dfrac{1}{(s-4)^2}$

12. $\mathscr{L}\left\{e^t\cos^2 3t\right\}=\mathscr{L}\left\{\dfrac{1}{2}e^t+\dfrac{1}{2}e^t\cos 6t\right\}=\dfrac{1}{2}\dfrac{1}{s-1}+\dfrac{1}{2}\dfrac{s-1}{(s-1)^2+36}$

15. $\mathscr{L}^{-1}\left\{\dfrac{1}{s^2-6s+10}\right\}=\mathscr{L}^{-1}\left\{\dfrac{1}{(s-3)^2+1^2}\right\}=e^{3t}\sin t$

18. $\mathscr{L}^{-1}\left\{\dfrac{2s+5}{s^2+6s+34}\right\}=\mathscr{L}^{-1}\left\{2\dfrac{(s+3)}{(s+3)^2+5^2}-\dfrac{1}{5}\dfrac{5}{(s+3)^2+5^2}\right\}=2e^{-3t}\cos 5t-\dfrac{1}{5}e^{-3t}\sin 5t$

21. $\mathscr{L}^{-1}\left\{\dfrac{2s-1}{s^2(s+1)^3}\right\}=\mathscr{L}^{-1}\left\{\dfrac{5}{s}-\dfrac{1}{s^2}-\dfrac{5}{s+1}-\dfrac{4}{(s+1)^2}-\dfrac{3}{2}\dfrac{2}{(s+1)^3}\right\}=5-t-5e^{-t}-4te^{-t}-\dfrac{3}{2}t^2e^{-t}$

24. $\mathscr{L}\{e^{2-t}\mathscr{U}(t-2)\}=\mathscr{L}\left\{e^{-(t-2)}\mathscr{U}(t-2)\right\}=\dfrac{e^{-2s}}{s+1}$

27. $\mathscr{L}\{\cos 2t\,\mathscr{U}(t-\pi)\}=\mathscr{L}\{\cos 2(t-\pi)\,\mathscr{U}(t-\pi)\}=\dfrac{se^{-\pi s}}{s^2+4}$

30. $\mathscr{L}\left\{te^{t-5}\mathscr{U}(t-5)\right\}=\mathscr{L}\left\{(t-5)e^{t-5}\mathscr{U}(t-5)+5e^{t-5}\mathscr{U}(t-5)\right\}=\dfrac{e^{-5s}}{(s-1)^2}+\dfrac{5e^{-5s}}{s-1}$

33. $\mathscr{L}^{-1}\left\{\dfrac{e^{-\pi s}}{s^2+1}\right\}=\sin(t-\pi)\,\mathscr{U}(t-\pi)$

36. $\mathscr{L}^{-1}\left\{\dfrac{e^{-2s}}{s^2(s-1)}\right\}=\mathscr{L}^{-1}\left\{-\dfrac{e^{-2s}}{s}-\dfrac{e^{-2s}}{s^2}+\dfrac{e^{-2s}}{s-1}\right\}=-\mathscr{U}(t-2)-(t-2)\,\mathscr{U}(t-2)+e^{t-2}\,\mathscr{U}(t-2)$

39. $\mathscr{L}\{t^2\sinh t\}=\dfrac{d^2}{ds^2}\left(\dfrac{1}{s^2-1}\right)=\dfrac{6s^2+2}{(s^2-1)^3}$

42. $\mathscr{L}\left\{te^{-3t}\cos 3t\right\} = -\dfrac{d}{ds}\left(\dfrac{s+3}{(s+3)^2+9}\right) = \dfrac{(s+3)^2-9}{[(s+3)^2+9]^2}$

45. (c)

48. (b)

51. $\mathscr{L}\{2-4\mathscr{U}(t-3)\} = \dfrac{2}{s} - \dfrac{4}{s}e^{-3s}$

54. $\mathscr{L}\left\{\sin t\,\mathscr{U}\left(t-\dfrac{3\pi}{2}\right)\right\} = \mathscr{L}\left\{-\cos\left(t-\dfrac{3\pi}{2}\right)\mathscr{U}\left(t-\dfrac{3\pi}{2}\right)\right\} = -\dfrac{se^{-3\pi s/2}}{s^2+1}$

57. $\mathscr{L}\{f(t)\} = \mathscr{L}\{\mathscr{U}(t-a) - \mathscr{U}(t-b)\} = \dfrac{e^{-as}}{s} - \dfrac{e^{-bs}}{s}$

60. $\mathscr{L}^{-1}\left\{\dfrac{2}{s} - \dfrac{3e^{-s}}{s^2} + \dfrac{5e^{-2s}}{s^2}\right\} = 2 - 3(t-1)\mathscr{U}(t-1) + 5(t-2)\mathscr{U}(t-2)$

$$= \begin{cases} 2, & 0 \le t < 1 \\ -3t+5, & 1 \le t < 2 \\ 2t-5, & t \ge 2 \end{cases}$$

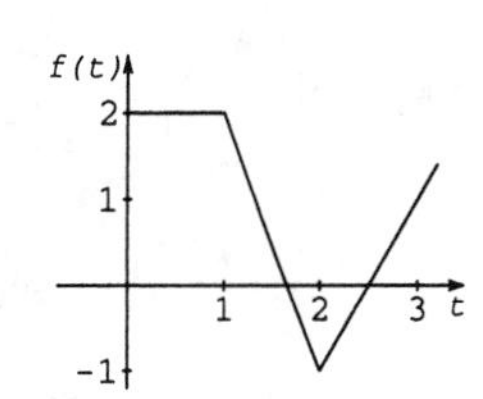

63. Since

$$t^2 - 3t = t(t-3) = (t-2+2)(t-2-1) = (t-2)^2 + (t-2) - 2$$

we have

$$\begin{aligned}\mathscr{L}\{(t^2-3t)\mathscr{U}(t-2)\} &= \mathscr{L}\{(t-2)^2\mathscr{U}(t-2) + (t-2)\mathscr{U}(t-2) - 2\mathscr{U}(t-2)\} \\ &= \frac{2}{s^3}e^{-2s} + \frac{1}{s^2}e^{-2s} - \frac{2}{s}e^{-2s}.\end{aligned}$$

Using the alternative form of the second translation theorem we obtain

$$\begin{aligned}\mathscr{L}\{(t^2-3t)\mathscr{U}(t-2)\} &= e^{-2s}\mathscr{L}\{(t+2)^2 - 3(t+2)\} \\ &= e^{-2s}\mathscr{L}\{t^2+t-2\} = e^{-2s}\left(\frac{2}{s^3} + \frac{1}{s^2} - \frac{2}{s}\right).\end{aligned}$$

Exercises 7.4

3. $\mathscr{L}\{y''+3y'\} = \mathscr{L}\{y''\} + 3\mathscr{L}\{y'\} = s^2Y(s) - sy(0) - y'(0) + 3[sY(s) - y(0)] = (s^2+3s)Y(s) - s - 2$

Exercises 7.4

6. We solve $\mathscr{L}\{y''+y\} = \mathscr{L}\{1\} = 1/s$.

$$s^2Y(s) - sy(0) - y'(0) + Y(s) = \frac{1}{s}$$

$$(s^2+1)Y(s) - 2s - 3 = \frac{1}{s}$$

$$Y(s) = \frac{1}{s(s^2+1)} + \frac{2s+3}{s^2+1}$$

9. $\mathscr{L}\left\{\int_0^t e^{-\tau}\cos\tau\,d\tau\right\} = \frac{1}{s}\mathscr{L}\{e^{-t}\cos t\} = \frac{1}{s}\frac{s+1}{(s+1)^2+1} = \frac{s+1}{s(s^2+2s+2)}$

12. $\mathscr{L}\left\{\int_0^t \sin\tau\cos(t-\tau)\,d\tau\right\} = \mathscr{L}\{\sin t\}\,\mathscr{L}\{\cos t\} = \frac{s}{(s^2+1)^2}$

15. $\mathscr{L}\{1 * t^3\} = \frac{1}{s}\frac{3!}{s^4} = \frac{6}{s^5}$

18. $\mathscr{L}\{t^2 * te^t\} = \frac{2}{s^3(s-1)^2}$

21. $\mathscr{L}^{-1}\left\{\frac{1}{s+5}F(s)\right\} = e^{-5t} * f(t) = \int_0^t f(\tau)e^{-5(t-\tau)}\,d\tau$

24. We use repeated applications of $\int_0^t f(\tau)\,d\tau = \mathscr{L}^{-1}\{F(s)/s\}$.

$$\mathscr{L}^{-1}\left\{\frac{1}{s(s-1)}\right\} = \int_0^t e^{\tau}\,d\tau = e^t - 1$$

$$\mathscr{L}^{-1}\left\{\frac{1}{s^2(s-1)}\right\} = \int_0^t (e^{\tau}-1)\,d\tau = e^t - t - 1$$

$$\mathscr{L}^{-1}\left\{\frac{1}{s^3(s-1)}\right\} = \int_0^t (e^{\tau}-\tau-1)\,d\tau = e^t - \frac{1}{2}t^2 - t - 1$$

27. $\mathscr{L}^{-1}\left\{\frac{s}{(s^2+4)^2}\right\} = \cos 2t * \frac{1}{2}\sin 2t = \frac{1}{2}\int_0^t \cos 2\tau \sin 2(t-\tau)\,d\tau$

$$= \frac{1}{2}\int_0^t \cos 2\tau(\sin 2t\cos 2\tau - \cos 2t\sin 2\tau)\,d\tau = \frac{1}{2}\left[\sin 2t\int_0^t \cos^2 2\tau\,d\tau - \cos 2t\int_0^t \frac{1}{2}\sin 4\tau\,d\tau\right]$$

$$= \frac{1}{2}\sin 2t\left[\frac{1}{2}\tau + \frac{1}{8}\sin 4\tau\right]_0^t - \frac{1}{4}\cos 2t\left[-\frac{1}{4}\cos 4\tau\right]_0^t$$

$$= \frac{1}{2}\sin 2t\left(\frac{1}{2}t + \frac{1}{8}\sin 4t\right) + \frac{1}{16}\cos 2t(\cos 4t - 1)$$

$$= \frac{1}{4}t\sin 2t + \frac{1}{16}\sin 2t\sin 4t + \frac{1}{16}\cos 2t\cos 4t - \frac{1}{16}\cos 2t$$

$$= \frac{1}{4}t\sin 2t + \frac{1}{16}\left[\sin 2t(2\sin 2t\cos 2t) + \cos 2t\left(\cos^2 2t - \sin^2 2t\right) - \cos 2t\right]$$

$$= \frac{1}{4}t\sin 2t + \frac{1}{16}\cos 2t\left[2\sin^2 2t + \cos^2 2t - \sin^2 2t - 1\right] = \frac{1}{4}t\sin 2t$$

30. $f*(g+h) = \int_0^t f(\tau)[g(t-\tau)+h(t-\tau)]\,d\tau = \int_0^t f(\tau)g(t-\tau)\,d\tau + \int_0^t f(\tau)h(t-\tau)\,d\tau$

$$= \int_0^t f(\tau)[g(t-\tau)+h(t-\tau)]\,d\tau = f*g + f*h$$

33. $\mathscr{L}\{f(t)\} = \dfrac{1}{1-e^{-bs}}\displaystyle\int_0^b \frac{a}{b}te^{-st}dt = \frac{a}{s}\left(\frac{1}{bs} - \frac{1}{e^{bs}-1}\right)$

36. $\mathscr{L}\{f(t)\} = \dfrac{1}{1-e^{-2\pi s}}\displaystyle\int_0^\pi e^{-st}\sin t\,dt = \frac{1}{s^2+1}\cdot\frac{1}{1-e^{-\pi s}}$

Exercises 7.5

3. The Laplace transform of the differential equation is

$$s\,\mathscr{L}\{y\} - y(0) + 4\,\mathscr{L}\{y\} = \frac{1}{s+4}.$$

Solving for $\mathscr{L}\{y\}$ we obtain $\mathscr{L}\{y\} = \dfrac{1}{(s+4)^2} + \dfrac{2}{s+4}$.

Thus
$$y = te^{-4t} + 2e^{-4t}.$$

6. The Laplace transform of the differential equation is

$$s^2\,\mathscr{L}\{y\} - sy(0) - y'(0) - 6\left[s\,\mathscr{L}\{y\} - y(0)\right] + 13\,\mathscr{L}\{y\} = 0.$$

Solving for $\mathscr{L}\{y\}$ we obtain

$$\mathscr{L}\{y\} = -\frac{3}{s^2-6s+13} = -\frac{3}{2}\frac{2}{(s-3)^2+2^2}.$$

Thus
$$y = -\frac{3}{2}e^{3t}\sin 2t.$$

9. The Laplace transform of the differential equation is

$$s^2\,\mathscr{L}\{y\} - sy(0) - y'(0) - 4\left[s\,\mathscr{L}\{y\} - y(0)\right] + 4\,\mathscr{L}\{y\} = \frac{6}{(s-2)^4}.$$

Solving for $\mathscr{L}\{y\}$ we obtain $\mathscr{L}\{y\} = \dfrac{1}{20}\dfrac{5!}{(s-2)^6}$. Thus, $y = \dfrac{1}{20}t^5e^{2t}$.

12. The Laplace transform of the differential equation is

$$s^2\,\mathscr{L}\{y\} - sy(0) - y'(0) + 16\,\mathscr{L}\{y\} = \frac{1}{s}.$$

Solving for $\mathscr{L}\{y\}$ we obtain

$$\mathscr{L}\{y\} = \frac{s^2+2s+1}{s(s^2+16)} = \frac{1}{16}\frac{1}{s} + \frac{15}{16}\frac{s}{s^2+4^2} + \frac{1}{2}\frac{4}{s^2+4^2}.$$

Thus

$$y = \frac{1}{16} + \frac{15}{16}\cos 4t + \frac{1}{2}\sin 4t.$$

15. The Laplace transform of the differential equation is

$$2\left[s^3\,\mathscr{L}\{y\} - s^2(0) - sy'(0) - y''(0)\right] + 3[s^2\,\mathscr{L}\{y\} - sy(0) - y'(0)] - 3[s\,\mathscr{L}\{y\} - y(0)] - 2\,\mathscr{L}\{y\} = \frac{1}{s+1}.$$

Solving for $\mathscr{L}\{y\}$ we obtain

$$\mathscr{L}\{y\} = \frac{2s+3}{(s+1)(s-1)(2s+1)(s+2)} = \frac{1}{2}\frac{1}{s+1} + \frac{5}{18}\frac{1}{s-1} - \frac{8}{9}\frac{1}{s+1/2} + \frac{1}{9}\frac{1}{s+2}.$$

Thus

$$y = \frac{1}{2}e^{-t} + \frac{5}{18}e^{t} - \frac{8}{9}e^{-t/2} + \frac{1}{9}e^{-2t}.$$

18. The Laplace transform of the differential equation is

$$s^4\,\mathscr{L}\{y\} - s^3y(0) - s^2y'(0) - sy''(0) - y'''(0) - \mathscr{L}\{y\} = \frac{1}{s^2}.$$

Solving for $\mathscr{L}\{y\}$ we obtain

$$\mathscr{L}\{y\} = \frac{1}{s^2(s^4-1)} = -\frac{1}{s^2} + \frac{1}{4}\frac{1}{s-1} - \frac{1}{4}\frac{1}{s+1} + \frac{1}{2}\frac{1}{s^2+1}.$$

Thus

$$y = -t + \frac{1}{4}e^{t} - \frac{1}{4}e^{-t} + \frac{1}{2}\sin t.$$

21. The Laplace transform of the differential equation is

$$s\,\mathscr{L}\{y\} - y(0) + 2\,\mathscr{L}\{y\} = \frac{1}{s^2} - e^{-s}\frac{s+1}{s^2}.$$

Solving for $\mathscr{L}\{y\}$ we obtain

$$\mathscr{L}\{y\} = \frac{1}{s^2(s+2)} - e^{-s}\frac{s+1}{s^2(s+1)} = -\frac{1}{4}\frac{1}{s} + \frac{1}{2}\frac{1}{s^2} + \frac{1}{4}\frac{1}{s+2} - e^{-s}\left[\frac{1}{4}\frac{1}{s} + \frac{1}{2}\frac{1}{s^2} - \frac{1}{4}\frac{1}{s+2}\right].$$

Thus

$$y = -\frac{1}{4} + \frac{1}{2}t + \frac{1}{4}e^{-2t} - \left[\frac{1}{4} + \frac{1}{2}(t-1) - \frac{1}{4}e^{-2(t-1)}\right]\mathscr{U}(t-1).$$

24. The Laplace transform of the differential equation is

$$s^2\,\mathscr{L}\{y\} - sy(0) - y'(0) - 5\,[s\,\mathscr{L}\{y\} - y(0)] + 6\,\mathscr{L}\{y\} = \frac{e^{-s}}{s}.$$

Solving for $\mathscr{L}\{y\}$ we obtain

$$\mathscr{L}\{y\} = e^{-s}\frac{1}{s(s-2)(s-3)} + \frac{1}{(s-2)(s-3)}$$

$$= e^{-s}\left[\frac{1}{6}\frac{1}{s} - \frac{1}{2}\frac{1}{s-2} + \frac{1}{3}\frac{1}{s-3}\right] - \frac{1}{s-2} + \frac{1}{s-3}.$$

Thus
$$y = \left[\frac{1}{6} - \frac{1}{2}e^{2(t-1)} + \frac{1}{3}e^{3(t-1)}\right]\mathscr{U}(t-1) + e^{3t} - e^{2t}.$$

27. Taking the Laplace transform of both sides of the differential equation and letting $c = y(0)$ we obtain

$$\mathscr{L}\{y''\} + \mathscr{L}\{2y'\} + \mathscr{L}\{y\} = 0$$

$$s^2\mathscr{L}\{y\} - sy(0) - y'(0) + 2s\,\mathscr{L}\{y\} - 2y(0) + \mathscr{L}\{y\} = 0$$

$$s^2\mathscr{L}\{y\} - cs - 2 + 2s\,\mathscr{L}\{y\} - 2c + \mathscr{L}\{y\} = 0$$

$$\left(s^2 + 2s + 1\right)\mathscr{L}\{y\} = cs + 2c + 2$$

$$\mathscr{L}\{y\} = \frac{cs}{(s+1)^2} + \frac{2c+2}{(s+1)^2}$$

$$= c\frac{s+1-1}{(s+1)^2} + \frac{2c+2}{(s+1)^2}$$

$$= \frac{c}{s+1} + \frac{c+2}{(s+1)^2}.$$

Therefore,

$$y(t) = c\mathscr{L}^{-1}\left\{\frac{1}{s+1}\right\} + (c+2)\,\mathscr{L}^{-1}\left\{\frac{1}{(s+1)^2}\right\} = ce^{-t} + (c+2)te^{-t}.$$

To find c we let $y(1) = 2$. Then $2 = ce^{-1} + (c+2)e^{-1} = 2(c+1)e^{-1}$ and $c = e - 1$.

Thus
$$y(t) = (e-1)e^{-t} + (e+1)te^{-t}.$$

30. The Laplace transform of the given equation is

$$\mathscr{L}\{f\} = \mathscr{L}\{2t\} - 4\,\mathscr{L}\{\sin t\}\,\mathscr{L}\{f\}.$$

Solving for $\mathscr{L}\{f\}$ we obtain

$$\mathscr{L}\{f\} = \frac{2s^2+2}{s^2(s^2+5)} = \frac{2}{5}\frac{1}{s^2} + \frac{8}{5\sqrt{5}}\frac{\sqrt{5}}{s^2+5}.$$

Thus
$$f(t) = \frac{2}{5}t + \frac{8}{5\sqrt{5}}\sin\sqrt{5}\,t.$$

33. The Laplace transform of the given equation is

$$\mathscr{L}\{f\} + \mathscr{L}\{1\}\mathscr{L}\{f\} = \mathscr{L}\{1\}.$$

Solving for $\mathscr{L}\{f\}$ we obtain $\mathscr{L}\{f\} = \dfrac{1}{s+1}$. Thus, $f(t) = e^{-t}$.

36. The Laplace transform of the given equation is

$$\mathscr{L}\{t\} - 2\mathscr{L}\{f\} = \mathscr{L}\left\{e^t - e^{-t}\right\}\mathscr{L}\{f\}.$$

Solving for $\mathscr{L}\{f\}$ we obtain

$$\mathscr{L}\{f\} = \frac{s^2-1}{2s^4} = \frac{1}{2}\frac{1}{s^2} - \frac{1}{12}\frac{3!}{s^4}.$$

Thus

$$f(t) = \frac{1}{2}t - \frac{1}{12}t^3.$$

39. From equation (3) in the text the differential equation is

$$0.005\frac{di}{dt} + i + 50\int_0^t i(\tau)\,d\tau = 100[1 - \mathscr{U}(t-1)], \quad i(0) = 0.$$

The Laplace transform of this equation is

$$0.005[s\mathscr{L}\{i\} - i(0)] + \mathscr{L}\{i\} + 50\frac{1}{s}\mathscr{L}\{i\} = 100\left[\frac{1}{s} - \frac{1}{s}e^{-s}\right].$$

Solving for $\mathscr{L}\{i\}$ we obtain

$$\mathscr{L}\{i\} = \frac{20{,}000}{(s+100)^2}(1 - e^{-s}).$$

Thus

$$i(t) = 20{,}000te^{-100t} - 20{,}000(t-1)e^{-100(t-1)}\mathscr{U}(t-1).$$

42. The differential equation is

$$10\frac{dq}{dt} + 10q = 30e^t - 30e^t\mathscr{U}(t-1.5).$$

The Laplace transform of this equation is

$$s\mathscr{L}\{q\} - q_0 + \mathscr{L}\{q\} = \frac{3}{s-1} - \frac{3e^{1.5}}{s-1.5}e^{-1.5s}.$$

Solving for $\mathscr{L}\{q\}$ we obtain

$$\mathscr{L}\{q\} = \left(q_0 - \frac{3}{2}\right)\cdot\frac{1}{s+1} + \frac{3}{2}\cdot\frac{1}{s-1} 3e^{1.5}\left(\frac{-2/5}{s+1} + \frac{2/5}{s-1.5}\right)e^{-1.55}..$$

Thus

$$q(t) = \left(q_0 - \frac{3}{2}\right)e^{-t} + \frac{3}{2}e^t + \frac{6}{5}e^{1.5}\left(e^{-(t-1.5)} - e^{1.5(t-1.5)}\right)\mathscr{U}(t-1.5).$$

45. **(a)** The differential equation is

$$\frac{di}{dt} + 10i = \sin t + \cos\left(t - \frac{3\pi}{2}\right)\mathcal{U}\left(t - \frac{3\pi}{2}\right), \quad i(0) = 0.$$

The Laplace transform of this equation is

$$s\mathscr{L}\{i\} + 10\mathscr{L}\{i\} = \frac{1}{s^2+1} + \frac{se^{-3\pi s/2}}{s^2+1}.$$

Solving for $\mathscr{L}\{i\}$ we obtain

$$\mathscr{L}\{i\} = \frac{1}{(s^2+1)(s+10)} + \frac{s}{(s^2+1)(s+10)}e^{-3\pi s/2}$$

$$= \frac{1}{101}\left(\frac{1}{s+10} - \frac{s}{s^2+1} + \frac{10}{s^2+1}\right) + \frac{1}{101}\left(\frac{-10}{s+10} + \frac{10s}{s^2+1} + \frac{1}{s^2+1}\right)e^{-3\pi s/2}.$$

Thus

$$i(t) = \frac{1}{101}\left(e^{-10t} - \cos t + 10\sin t\right)$$

$$+ \frac{1}{101}\left[-10e^{-10(t-3\pi/2)} + 10\cos\left(t - \frac{3\pi}{2}\right) + \sin\left(t - \frac{3\pi}{2}\right)\right]\mathcal{U}\left(t - \frac{3\pi}{2}\right).$$

(b)

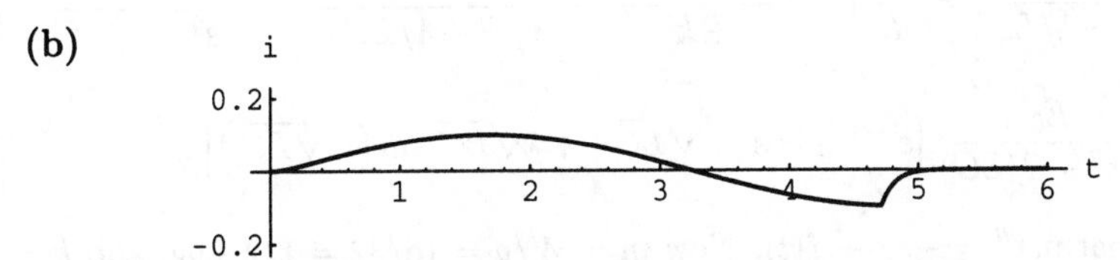

The maximum value of $i(t)$ is approximately 0.1 at $t = 1.7$, the minimum is approximately -0.1 at 4.7.

48. The differential equation is

$$\frac{d^2q}{dt^2} + 20\frac{dq}{dt} + 200q = 150, \quad q(0) = q'(0) = 0.$$

The Laplace transform of this equation is

$$s^2\mathscr{L}\{q\} + 20s\mathscr{L}\{q\} + 200\mathscr{L}\{q\} = \frac{150}{s}.$$

Solving for $\mathscr{L}\{q\}$ we obtain

$$\mathscr{L}\{q\} = \frac{150}{s(s^2+20s+200)} = \frac{3}{4}\frac{1}{s} - \frac{3}{4}\frac{s+10}{(s+10)^2+10^2} - \frac{3}{4}\frac{10}{(s+10)^2+10^2}.$$

Thus

$$q(t) = \frac{3}{4} - \frac{3}{4}e^{-10t}\cos 10t - \frac{3}{4}e^{-10t}\sin 10t$$

and

$$i(t) = q'(t) = 15e^{-10t}\sin 10t.$$

If $E(t) = 150 - 150\mathscr{U}(t-2)$, then

$$\mathscr{L}\{q\} = \frac{150}{s(s^2+20s+200)}\left(1-e^{-2s}\right)$$

$$q(t) = \frac{3}{4} - \frac{3}{4}e^{-10t}\cos 10t - \frac{3}{4}e^{-10t}\sin 10t - \left[\frac{3}{4} - \frac{3}{4}e^{-10(t-2)}\cos 10(t-2)\right.$$

$$\left. - \frac{3}{4}e^{-10(t-2)}\sin 10(t-2)\right]\mathscr{U}(t-2).$$

51. The differential equation is

$$\frac{d^2q}{dt^2} + \frac{1}{LC}q = \frac{E_0}{L}e^{-kt}, \quad q(0) = q'(0) = 0.$$

The Laplace transform of this equation is

$$s^2\mathscr{L}\{q\} + \frac{1}{LC}\mathscr{L}\{q\} = \frac{E_0}{L}\frac{1}{s+k}.$$

Solving for $\mathscr{L}\{q\}$ we obtain

$$\mathscr{L}\{q\} = \frac{E_0}{L}\frac{1}{(s+k)(s^2+1/LC)} = \frac{E_0}{L}\left(\frac{1/(k^2+1/LC)}{s+k} - \frac{s/(k^2+1/LC)}{s^2+1/LC} + \frac{k/(k^2+1/LC)}{s^2+1/LC}\right).$$

Thus $$q(t) = \frac{E_0}{L(k^2+1/LC)}\left[e^{-kt} - \cos\left(t/\sqrt{LC}\right) + k\sqrt{LC}\sin\left(t/\sqrt{LC}\right)\right].$$

54. Recall from Chapter 5 that $mx'' = -kx + f(t)$. Now $m = W/g = 16/32 = 1/2$ slug, and $k = 4.5$, so the differential equation is

$$\frac{1}{2}x'' + 4.5x = 4\sin 3t + 2\cos 3t \quad \text{or} \quad x'' + 9x = 8\sin 3t + 4\cos 3t.$$

The initial conditions are $x(0) = x'(0) = 0$. The Laplace transform of the differential equation is

$$s^2\mathscr{L}\{x\} + 9\mathscr{L}\{x\} = \frac{24}{s^2+9} + \frac{4s}{s^2+9}.$$

Solving for $\mathscr{L}\{x\}$ we obtain

$$\mathscr{L}\{x\} = \frac{4s+24}{(s^2+9)^2} = \frac{2}{3}\frac{2(3)s}{(s^2+9)^2} + \frac{12}{27}\frac{2(3)^3}{(s^2+9)^2}.$$

Thus $$x(t) = \frac{2}{3}t\sin 3t + \frac{4}{9}(\sin 3t - 3t\cos 3t) = \frac{2}{3}t\sin 3t + \frac{4}{9}\sin 3t - \frac{4}{3}t\cos 3t.$$

57. The differential equation is

$$EI\frac{d^4y}{dx^4} = \frac{2w_0}{L}\left[\frac{L}{2} - x + \left(x - \frac{L}{2}\right)\mathscr{U}\left(x - \frac{L}{2}\right)\right].$$

Taking the Laplace transform of both sides and using $y(0) = y'(0) = 0$ we obtain

$$s^4\,\mathscr{L}\{y\} - sy''(0) - y'''(0) = \frac{2w_0}{EIL}\left[\frac{L}{2s} - \frac{1}{s^2} + \frac{1}{s^2}\,e^{-Ls/2}\right].$$

Letting $y''(0) = c_1$ and $y'''(0) = c_2$ we have

$$\mathscr{L}\{y\} = \frac{c_1}{s^3} + \frac{c_2}{s^4} + \frac{2w_0}{EIL}\left[\frac{L}{2s^5} - \frac{1}{s^6} + \frac{1}{s^6}\,e^{-Ls/2}\right]$$

so that

$$y(x) = \frac{1}{2}c_1x^2 + \frac{1}{6}c_2x^3 + \frac{2w_0}{EIL}\left[\frac{L}{48}x^4 - \frac{1}{120}x^5 + \frac{1}{120}\left(x - \frac{L}{2}\right)\mathscr{U}\left(x - \frac{L}{2}\right)\right]$$

$$= \frac{1}{2}c_1x^2 + \frac{1}{6}c_2x^3 + \frac{w_0}{60EIL}\left[\frac{5L}{2}x^4 - x^5 + \left(x - \frac{L}{2}\right)^5\mathscr{U}\left(x - \frac{L}{2}\right)\right].$$

To find c_1 and c_2 we compute

$$y''(x) = c_1 + c_2x + \frac{w_0}{60EIL}\left[30Lx^2 - 20x^3 + 20\left(x - \frac{L}{2}\right)^3\mathscr{U}\left(x - \frac{L}{2}\right)\right]$$

and

$$y'''(x) = c_2 + \frac{w_0}{60EIL}\left[60Lx - 60x^2 + 60\left(x - \frac{L}{2}\right)^2\mathscr{U}\left(x - \frac{L}{2}\right)\right].$$

Then $y''(L) = y'''(L) = 0$ yields the system

$$c_1 + c_2L + \frac{w_0}{60EIL}\left[30L^3 - 20L^3 + \frac{5}{2}L^3\right] = c_1 + c_2L + \frac{5w_0L^2}{24EI} = 0$$

$$c_2 + \frac{w_0}{60EIL}[60L^2 - 60L^2 + 15L^2] = c_2 + \frac{w_0L}{4EI} = 0.$$

Solving for c_1 and c_2 we obtain $c_1 = w_0L^2/24EI$ and $c_2 = -w_0L/4EI$. Thus

$$y(x) = \frac{w_0L^2}{48EI}x^2 - \frac{w_0L}{24EI} + \frac{w_0}{60EIL}\left[\frac{5L}{2}x^4 - x^5 + \left(x - \frac{L}{2}\right)^5\mathscr{U}\left(x - \frac{L}{2}\right)\right].$$

Exercises 7.6

3. The Laplace transform of the differential equation yields

$$\mathscr{L}\{y\} = \frac{1}{s^2+1}\left(1 + e^{-2\pi s}\right)$$

so that

$$y = \sin t + \sin t\,\mathscr{U}(t - 2\pi).$$

6. The Laplace transform of the differential equation yields

$$\mathscr{L}\{y\} = \frac{s}{s^2+1} + \frac{1}{s^2+1}(e^{-2\pi s} + e^{-4\pi s})$$

so that
$$y = \cos t + \sin t[\mathscr{U}(t-2\pi) + \mathscr{U}(t-4\pi)].$$

9. The Laplace transform of the differential equation yields
$$\mathscr{L}\{y\} = \frac{1}{(s+2)^2+1}e^{-2\pi s}$$
so that
$$y = e^{-2(t-2\pi)}\sin t\,\mathscr{U}(t-2\pi).$$

12. The Laplace transform of the differential equation yields
$$\mathscr{L}\{y\} = \frac{1}{(s-1)^2(s-6)} + \frac{e^{-2s}+e^{-4s}}{(s-1)(s-6)}$$
$$= -\frac{1}{25}\frac{1}{s-1} - \frac{1}{5}\frac{1}{(s-1)^2} + \frac{1}{25}\frac{1}{s-6} + \left[-\frac{1}{5}\frac{1}{s-1} + \frac{1}{5}\frac{1}{s-6}\right]\left(e^{-2s}+e^{-4s}\right)$$
so that
$$y = -\frac{1}{25}e^t - \frac{1}{5}te^t + \frac{1}{25}e^{6t} + \left[-\frac{1}{5}e^{t-2} + \frac{1}{5}e^{6(t-2)}\right]\mathscr{U}(t-2)$$
$$+ \left[-\frac{1}{5}e^{t-4} + \frac{1}{5}e^{6(t-4)}\right]\mathscr{U}(t-4).$$

Exercises 7.7

3. Taking the Laplace transform of the system gives
$$s\mathscr{L}\{x\} + 1 = \mathscr{L}\{x\} - 2\mathscr{L}\{y\}$$
$$s\mathscr{L}\{y\} - 2 = 5\mathscr{L}\{x\} - \mathscr{L}\{y\}$$
so that
$$\mathscr{L}\{x\} = \frac{-s-5}{s^2+9} = -\frac{s}{s^2+9} - \frac{5}{3}\frac{3}{s^2+9}$$
and
$$x = -\cos 3t - \frac{5}{3}\sin 3t.$$
Then
$$y = \frac{1}{2}x - \frac{1}{2}x' = 2\cos 3t - \frac{7}{3}\sin 3t.$$

6. Taking the Laplace transform of the system gives
$$(s+1)\mathscr{L}\{x\} - (s-1)\mathscr{L}\{y\} = -1$$
$$s\mathscr{L}\{x\} + (s+2)\mathscr{L}\{y\} = 1$$

so that $$\mathscr{L}\{y\} = \frac{s+1/2}{s^2+s+1} = \frac{s+1/2}{(s+1/2)^2+(\sqrt{3}/2)^2}$$

and $$\mathscr{L}\{x\} = \frac{-3/2}{s^2+s+1} = \frac{-3/2}{(s+1/2)^2+(\sqrt{3}/2)^2}.$$

Then $$y = e^{-t/2}\cos\frac{\sqrt{3}}{2}t \quad \text{and} \quad x = e^{-t/2}\sin\frac{\sqrt{3}}{2}t.$$

9. Adding the equations and then subtracting them gives

$$\frac{d^2x}{dt^2} = \frac{1}{2}t^2 + 2t$$

$$\frac{d^2y}{dt^2} = \frac{1}{2}t^2 - 2t.$$

Taking the Laplace transform of the system gives

and $$\mathscr{L}\{x\} = 8\frac{1}{s} + \frac{1}{24}\frac{4!}{s^5} + \frac{1}{3}\frac{3!}{s^4}$$

$$\mathscr{L}\{y\} = \frac{1}{24}\frac{4!}{s^5} - \frac{1}{3}\frac{3!}{s^4}$$

so that $$x = 8 + \frac{1}{24}t^4 + \frac{1}{3}t^3 \quad \text{and} \quad y = \frac{1}{24}t^4 - \frac{1}{3}t^3.$$

12. Taking the Laplace transform of the system gives

$$(s-4)\,\mathscr{L}\{x\} + 2\mathscr{L}\{y\} = \frac{2e^{-s}}{s}$$

$$-3\,\mathscr{L}\{x\} + (s+1)\,\mathscr{L}\{y\} = \frac{1}{2} + \frac{e^{-s}}{s}$$

so that $$\mathscr{L}\{x\} = \frac{-1/2}{(s-1)(s-2)} + e^{-s}\frac{1}{(s-1)(s-2)}$$

$$= \left[\frac{1}{2}\frac{1}{s-1} - \frac{1}{2}\frac{1}{s-2}\right] + e^{-s}\left[-\frac{1}{s-1} + \frac{1}{s-2}\right]$$

and $$\mathscr{L}\{y\} = \frac{e^{-s}}{s} + \frac{s/4-1}{(s-1)(s-2)} + e^{-s}\frac{-s/2+2}{(s-1)(s-2)}$$

$$= \frac{3}{4}\frac{1}{s-1} - \frac{1}{2}\frac{1}{s-2} + e^{-s}\left[\frac{1}{s} - \frac{3}{2}\frac{1}{s-1} + \frac{1}{s-2}\right].$$

Then $$x = \frac{1}{2}e^t - \frac{1}{2}e^{2t} + \left[-e^{t-1} + e^{2(t-1)}\right]\mathcal{U}(t-1)$$

and $$y = \frac{3}{4}e^t - \frac{1}{2}e^{2t} + \left[1 - \frac{3}{2}e^{t-1} + e^{2(t-1)}\right]\mathcal{U}(t-1).$$

15. **(a)** By Kirchoff's first law we have $i_1 = i_2 + i_3$. By Kirchoff's second law, on each loop we have $E(t) = Ri_1 + L_1 i_2'$ and $E(t) = Ri_1 + L_2 i_3'$ or $L_1 i_2' + Ri_2 + Ri_3 = E(t)$ and $L_2 i_3' + Ri_2 + Ri_3 = E(t)$.

(b) Taking the Laplace transform of the system

$$0.01i_2' + 5i_2 + 5i_3 = 100$$

$$0.0125i_3' + 5i_2 + 5i_3 = 100$$

gives $$(s + 500)\,\mathcal{L}\{i_2\} + 500\mathcal{L}\{i_3\} = \frac{10{,}000}{s}$$

$$400\mathcal{L}\{i_2\} + (s + 400)\,\mathcal{L}\{i_3\} = \frac{8{,}000}{s}$$

so that $$\mathcal{L}\{i_3\} = \frac{8{,}000}{s^2 + 900s} = \frac{80}{9}\frac{1}{s} - \frac{80}{9}\frac{1}{s + 900}.$$

Then

$$i_3 = \frac{80}{9} - \frac{80}{9}e^{-900t} \quad \text{and} \quad i_2 = 20 - 0.0025i_3' - i_3 = \frac{100}{9} - \frac{100}{9}e^{-900t}.$$

(c) $i_1 = i_2 + i_3 = 20 - 20e^{-900t}$

18. Taking the Laplace transform of the system

$$0.5i_1' + 50i_2 = 60$$

$$0.005i_2' + i_2 - i_1 = 0$$

gives $$s\,\mathcal{L}\{i_1\} + 100\,\mathcal{L}\{i_2\} = \frac{120}{s}$$

$$-200\,\mathcal{L}\{i_1\} + (s + 200)\,\mathcal{L}\{i_2\} = 0$$

so that

$$\mathcal{L}\{i_2\} = \frac{24{,}000}{s(s^2 + 200s + 20{,}000)} = \frac{6}{5}\frac{1}{s} - \frac{6}{5}\frac{s + 100}{(s + 100)^2 + 100^2} - \frac{6}{5}\frac{100}{(s + 100)^2 + 100^2}.$$

Then $$i_2 = \frac{6}{5} - \frac{6}{5}e^{-100t}\cos 100t - \frac{6}{5}e^{-100t}\sin 100t$$

and $$i_1 = 0.005i_2' + i_2 = \frac{6}{5} - \frac{6}{5}e^{-100t}\cos 100t.$$

21. Taking the Laplace transform of the system

$$4\theta_1'' + \theta_2'' + 8\theta_1 = 0$$

$$\theta_1'' + \theta_2'' + 2\theta_2 = 0$$

gives
$$4\left(s^2+2\right)\mathscr{L}\{\theta_1\} + s^2\mathscr{L}\{\theta_2\} = 3s$$

$$s^2\mathscr{L}\{\theta_1\} + \left(s^2+2\right)\mathscr{L}\{\theta_2\} = 0$$

so that
$$\left(3s^2+4\right)\left(s^2+4\right)\mathscr{L}\{\theta_2\} = -3s^3$$

or
$$\mathscr{L}\{\theta_2\} = \frac{1}{2}\frac{s}{s^2+4/3} - \frac{3}{2}\frac{s}{s^2+4}.$$

Then
$$\theta_2 = \frac{1}{2}\cos\frac{2}{\sqrt{3}}t - \frac{3}{2}\cos 2t \quad \text{and} \quad \theta_1'' = -\theta_2'' - 2\theta_2$$

so that
$$\theta_1 = \frac{1}{4}\cos\frac{2}{\sqrt{3}}t + \frac{3}{4}\cos 2t.$$

Chapter 7 Review Exercises

3. False; consider $f(t) = t^{-1/2}$.

6. False; consider $f(t) = 1$ and $g(t) = 1$.

9. $\mathscr{L}\{\sin 2t\} = \dfrac{2}{s^2+4}$

12. $\mathscr{L}\{\sin 2t\,\mathscr{U}(t-\pi)\} = \mathscr{L}\{\sin 2(t-\pi)\mathscr{U}(t-\pi)\} = \dfrac{2}{s^2+4}e^{-\pi s}$

15. $\mathscr{L}^{-1}\left\{\dfrac{1}{(s-5)^3}\right\} = \mathscr{L}^{-1}\left\{\dfrac{1}{2}\dfrac{2}{(s-5)^3}\right\} = \dfrac{1}{2}t^2e^{5t}$

18. $\mathscr{L}^{-1}\left\{\dfrac{1}{s^2}e^{-5s}\right\} = (t-5)\mathscr{U}(t-5)$

21. $\mathscr{L}\left\{e^{-5t}\right\}$ exists for $s > -5$.

24. $1 * 1 = \displaystyle\int_0^t d\tau = t$

27. (a) $f(t) = 2 - 2\mathscr{U}(t-2) + [(t-2)+2]\mathscr{U}(t-2) = 2 + (t-2)\mathscr{U}(t-2)$

(b) $\mathscr{L}\{f(t)\} = \dfrac{2}{s} + \dfrac{1}{s^2}e^{-2s}$

(c) $\mathscr{L}\left\{e^t f(t)\right\} = \dfrac{2}{s-1} + \dfrac{1}{(s-1)^2}e^{-2(s-1)}$

30. Taking the Laplace transform of the differential equation we obtain

$$\mathscr{L}\{y\} = \frac{1}{(s-1)^2(s^2-8s+20)}$$

$$= \frac{6}{169}\frac{1}{s-1} + \frac{1}{13}\frac{1}{(s-1)^2} - \frac{6}{169}\frac{s-4}{(s-4)^2+2^2} + \frac{5}{338}\frac{2}{(s-4)^2+2^2}$$

so that
$$y = \frac{6}{169}e^t + \frac{1}{13}te^t - \frac{6}{169}e^{4t}\cos 2t + \frac{5}{338}e^{4t}\sin 2t.$$

33. Taking the Laplace transform of the differential equation we obtain

$$\mathscr{L}\{y\} = \frac{s^3+2}{s^3(s-5)} - \frac{2+2s+s^2}{s^3(s-5)}e^{-s}$$

$$= -\frac{2}{125}\frac{1}{s} - \frac{2}{25}\frac{1}{s^2} - \frac{1}{5}\frac{2}{s^3} + \frac{127}{125}\frac{1}{s-5} - \left[-\frac{37}{125}\frac{1}{s} - \frac{12}{25}\frac{1}{s^2} - \frac{1}{5}\frac{2}{s^3} + \frac{37}{125}\frac{1}{s-5}\right]e^{-s}$$

so that

$$y = -\frac{2}{125} - \frac{2}{25}t - \frac{1}{5}t^2 + \frac{127}{125}e^{5t} - \left[-\frac{37}{125} - \frac{12}{25}(t-1) - \frac{1}{5}(t-1)^2 + \frac{37}{125}e^{5(t-1)}\right]\mathscr{U}(t-1).$$

36. Taking the Laplace transform of the integral equation we obtain

$$(\mathscr{L}\{f\})^2 = 6\cdot\frac{6}{s^4} \quad \text{or} \quad \mathscr{L}\{f\} = \pm 6\cdot\frac{1}{s^2}$$

so that $f(t) = \pm 6t$.

39. The integral equation is

$$10i + 2\int_0^t i(\tau)\,d\tau = 2t^2 + 2t.$$

Taking the Laplace transform we obtain

$$\mathscr{L}\{i\} = \left(\frac{4}{s^3} + \frac{2}{s^2}\right)\frac{s}{10s+2} = \frac{s+2}{s^2(5s+2)} = -\frac{9}{s} + \frac{2}{s^2} + \frac{45}{5s+1} = -\frac{9}{s} + \frac{2}{s^2} + \frac{9}{s+1/5}.$$

Thus
$$i(t) = -9 + 2t + 9e^{-t/5}.$$

42. Taking the Laplace transform of the given differential equation we obtain

$$\mathscr{L}\{y\} = \frac{c_1}{2}\cdot\frac{2s}{s^4+4} + \frac{c_2}{4}\cdot\frac{4}{s^4+4} + \frac{w_0}{4EI}\cdot\frac{4}{s^4+4}e^{-s\pi/2}$$

so that

$$y = \frac{c_1}{2}\sin x \sinh x + \frac{c_2}{4}(\sin x \cosh x - \cos x \sinh x)$$

$$+ \frac{w_0}{4EI}\left[\sin\left(x - \frac{\pi}{2}\right)\cosh\left(x - \frac{\pi}{2}\right) - \cos\left(x - \frac{\pi}{2}\right)\sinh\left(x - \frac{\pi}{2}\right)\right]\mathcal{U}\left(x - \frac{\pi}{2}\right)$$

where $y''(0) = c_1$ and $y'''(0) = c_2$. Using $y(\pi) = 0$ and $y'(\pi) = 0$ we find

$$c_1 = \frac{w_0}{EI}\frac{\sinh\frac{\pi}{2}}{\sinh\pi}, \qquad c_2 = -\frac{w_0}{EI}\frac{\cosh\frac{\pi}{2}}{\sinh\pi}.$$

Hence

$$y = \frac{w_0}{2EI}\frac{\sinh\frac{\pi}{2}}{\sinh\pi}\sin x \sinh x - \frac{w_0}{4EI}\frac{\cosh\frac{\pi}{2}}{\sinh\pi}(\sin x \cosh x - \cos x \sinh x)$$

$$+ \frac{w_0}{4EI}\left[\sin\left(x - \frac{\pi}{2}\right)\cosh\left(x - \frac{\pi}{2}\right) - \cos\left(x - \frac{\pi}{2}\right)\sinh\left(x - \frac{\pi}{2}\right)\right]\mathcal{U}\left(x - \frac{\pi}{2}\right).$$

8 Systems of Linear First-Order Differential Equations

Exercises 8.1

3. Let $\mathbf{X} = \begin{pmatrix} x \\ y \\ z \end{pmatrix}$. Then

$$\mathbf{X}' = \begin{pmatrix} -3 & 4 & -9 \\ 6 & -1 & 0 \\ 10 & 4 & 3 \end{pmatrix} \mathbf{X}.$$

6. Let $\mathbf{X} = \begin{pmatrix} x \\ y \end{pmatrix}$. Then

$$\mathbf{X}' = \begin{pmatrix} -3 & 4 \\ 5 & 9 \end{pmatrix} \mathbf{X} + \begin{pmatrix} e^{-t} \sin 2t \\ 4e^{-t} \cos 2t \end{pmatrix}.$$

9. $\dfrac{dx}{dt} = x - y + 2z + e^{-t} - 3t;\quad \dfrac{dy}{dt} = 3x - 4y + z + 2e^{-t} + t;\quad \dfrac{dz}{dt} = -2x + 5y + 6z + 2e^{-t} - t$

12. Since

$$\mathbf{X}' = \begin{pmatrix} 5\cos t - 5\sin t \\ 2\cos t - 4\sin t \end{pmatrix} e^t \quad \text{and} \quad \begin{pmatrix} -2 & 5 \\ -2 & 4 \end{pmatrix} \mathbf{X} = \begin{pmatrix} 5\cos t - 5\sin t \\ 2\cos t - 4\sin t \end{pmatrix} e^t$$

we see that

$$\mathbf{X}' = \begin{pmatrix} -2 & 5 \\ -2 & 4 \end{pmatrix} \mathbf{X}.$$

15. Since

$$\mathbf{X}' = \begin{pmatrix} 0 \\ 0 \\ 0 \end{pmatrix} \quad \text{and} \quad \begin{pmatrix} 1 & 2 & 1 \\ 6 & -1 & 0 \\ -1 & -2 & -1 \end{pmatrix} \mathbf{X} = \begin{pmatrix} 0 \\ 0 \\ 0 \end{pmatrix}$$

we see that

$$\mathbf{X}' = \begin{pmatrix} 1 & 2 & 1 \\ 6 & -1 & 0 \\ -1 & -2 & -1 \end{pmatrix} \mathbf{X}.$$

18. Yes, since $W(\mathbf{X}_1, \mathbf{X}_2) = 8e^{2t} \neq 0$ and $\mathbf{X}_1$ and $\mathbf{X}_2$ are linearly independent on $-\infty < t < \infty$.

21. Since

$$\mathbf{X}_p' = \begin{pmatrix} 2 \\ -1 \end{pmatrix} \quad \text{and} \quad \begin{pmatrix} 1 & 4 \\ 3 & 2 \end{pmatrix} \mathbf{X}_p + \begin{pmatrix} 2 \\ -4 \end{pmatrix} t + \begin{pmatrix} -7 \\ -18 \end{pmatrix} = \begin{pmatrix} 2 \\ -1 \end{pmatrix}$$

we see that
$$\mathbf{X}_p' = \begin{pmatrix} 1 & 4 \\ 3 & 2 \end{pmatrix}\mathbf{X}_p + \begin{pmatrix} 2 \\ -4 \end{pmatrix}t + \begin{pmatrix} -7 \\ -18 \end{pmatrix}.$$

24. Since
$$\mathbf{X}_p' = \begin{pmatrix} 3\cos 3t \\ 0 \\ -3\sin 3t \end{pmatrix} \quad \text{and} \quad \begin{pmatrix} 1 & 2 & 3 \\ -4 & 2 & 0 \\ -6 & 1 & 0 \end{pmatrix}\mathbf{X}_p + \begin{pmatrix} -1 \\ 4 \\ 3 \end{pmatrix}\sin 3t = \begin{pmatrix} 3\cos 3t \\ 0 \\ -3\sin 3t \end{pmatrix}$$

we see that
$$\mathbf{X}_p' = \begin{pmatrix} 1 & 2 & 3 \\ -4 & 2 & 0 \\ -6 & 1 & 0 \end{pmatrix}\mathbf{X}_p + \begin{pmatrix} -1 \\ 4 \\ 3 \end{pmatrix}\sin 3t.$$

Exercises 8.2

3. The system is
$$\mathbf{X}' = \begin{pmatrix} -4 & 2 \\ -5/2 & 2 \end{pmatrix}\mathbf{X}$$
and $\det(\mathbf{A} - \lambda\mathbf{I}) = (\lambda - 1)(\lambda + 3) = 0$. For $\lambda_1 = 1$ we obtain
$$\left(\begin{array}{cc|c} -5 & 2 & 0 \\ -5/2 & 1 & 0 \end{array}\right) \Longrightarrow \left(\begin{array}{cc|c} -5 & 2 & 0 \\ 0 & 0 & 0 \end{array}\right) \quad \text{so that} \quad \mathbf{K}_1 = \begin{pmatrix} 2 \\ 5 \end{pmatrix}.$$
For $\lambda_2 = -3$ we obtain
$$\left(\begin{array}{cc|c} -1 & 2 & 0 \\ -5/2 & 5 & 0 \end{array}\right) \Longrightarrow \left(\begin{array}{cc|c} -1 & 2 & 0 \\ 0 & 0 & 0 \end{array}\right) \quad \text{so that} \quad \mathbf{K}_2 = \begin{pmatrix} 2 \\ 1 \end{pmatrix}.$$

Then
$$\mathbf{X} = c_1\begin{pmatrix} 2 \\ 5 \end{pmatrix}e^t + c_2\begin{pmatrix} 2 \\ 1 \end{pmatrix}e^{-3t}.$$

6. The system is
$$\mathbf{X}' = \begin{pmatrix} -6 & 2 \\ -3 & 1 \end{pmatrix}\mathbf{X}$$
and $\det(\mathbf{A} - \lambda\mathbf{I}) = \lambda(\lambda + 5) = 0$. For $\lambda_1 = 0$ we obtain
$$\left(\begin{array}{cc|c} -6 & 2 & 0 \\ -3 & 1 & 0 \end{array}\right) \Longrightarrow \left(\begin{array}{cc|c} 1 & -1/3 & 0 \\ 0 & 0 & 0 \end{array}\right) \quad \text{so that} \quad \mathbf{K}_1 = \begin{pmatrix} 1 \\ 3 \end{pmatrix}.$$
For $\lambda_2 = -5$ we obtain
$$\left(\begin{array}{cc|c} -1 & 2 & 0 \\ -3 & 6 & 0 \end{array}\right) \Longrightarrow \left(\begin{array}{cc|c} 1 & -2 & 0 \\ 0 & 0 & 0 \end{array}\right) \quad \text{so that} \quad \mathbf{K}_2 = \begin{pmatrix} 2 \\ 1 \end{pmatrix}.$$

Then $$\mathbf{X} = c_1 \begin{pmatrix} 1 \\ 3 \end{pmatrix} + c_2 \begin{pmatrix} 2 \\ 1 \end{pmatrix} e^{-5t}.$$

9. We have $\det(\mathbf{A} - \lambda\mathbf{I}) = -(\lambda+1)(\lambda-3)(\lambda+2) = 0$. For $\lambda_1 = -1$, $\lambda_2 = 3$, and $\lambda_3 = -2$ we obtain

$$\mathbf{K}_1 = \begin{pmatrix} -1 \\ 0 \\ 1 \end{pmatrix}, \quad \mathbf{K}_2 = \begin{pmatrix} 1 \\ 4 \\ 3 \end{pmatrix}, \quad \text{and} \quad \mathbf{K}_3 = \begin{pmatrix} 1 \\ -1 \\ 3 \end{pmatrix},$$

so that $$\mathbf{X} = c_1 \begin{pmatrix} -1 \\ 0 \\ 1 \end{pmatrix} e^{-t} + c_2 \begin{pmatrix} 1 \\ 4 \\ 3 \end{pmatrix} e^{3t} + c_3 \begin{pmatrix} 1 \\ -1 \\ 3 \end{pmatrix} e^{-2t}.$$

12. We have $\det(\mathbf{A} - \lambda\mathbf{I}) = (\lambda-3)(\lambda+5)(6-\lambda) = 0$. For $\lambda_1 = 3$, $\lambda_2 = -5$, and $\lambda_3 = 6$ we obtain

$$\mathbf{K}_1 = \begin{pmatrix} 1 \\ 1 \\ 0 \end{pmatrix}, \quad \mathbf{K}_2 = \begin{pmatrix} 1 \\ -1 \\ 0 \end{pmatrix}, \quad \text{and} \quad \mathbf{K}_3 = \begin{pmatrix} 2 \\ -2 \\ 11 \end{pmatrix},$$

so that $$\mathbf{X} = c_1 \begin{pmatrix} 1 \\ 1 \\ 0 \end{pmatrix} e^{3t} + c_2 \begin{pmatrix} 1 \\ -1 \\ 0 \end{pmatrix} e^{-5t} + c_3 \begin{pmatrix} 2 \\ -2 \\ 11 \end{pmatrix} e^{6t}.$$

15. $$\mathbf{X} = c_1 \begin{pmatrix} 0.382175 \\ 0.851161 \\ 0.359815 \end{pmatrix} e^{8.58979t} + c_2 \begin{pmatrix} 0.405188 \\ -0.676043 \\ 0.615458 \end{pmatrix} e^{2.25684t} + c_3 \begin{pmatrix} -0.923562 \\ -0.132174 \\ 0.35995 \end{pmatrix} e^{-0.0466321t}$$

18. We have $\det(\mathbf{A} - \lambda\mathbf{I}) = (\lambda+1)^2 = 0$. For $\lambda_1 = -1$ we obtain

$$\mathbf{K} = \begin{pmatrix} 1 \\ 1 \end{pmatrix}.$$

A solution of $(\mathbf{A} - \lambda_1\mathbf{I})\mathbf{P} = \mathbf{K}$ is

$$\mathbf{P} = \begin{pmatrix} 0 \\ 1/5 \end{pmatrix}$$

so that $$\mathbf{X} = c_1 \begin{pmatrix} 1 \\ 1 \end{pmatrix} e^{-t} + c_2 \left[\begin{pmatrix} 1 \\ 1 \end{pmatrix} te^{-t} + \begin{pmatrix} 0 \\ 1/5 \end{pmatrix} e^{-t} \right].$$

21. We have $\det(\mathbf{A} - \lambda\mathbf{I}) = (1-\lambda)(\lambda-2)^2 = 0$. For $\lambda_1 = 1$ we obtain

$$\mathbf{K}_1 = \begin{pmatrix} 1 \\ 1 \\ 1 \end{pmatrix}.$$

For $\lambda_2 = 2$ we obtain

$$\mathbf{K}_2 = \begin{pmatrix} 1 \\ 0 \\ 1 \end{pmatrix} \quad \text{and} \quad \mathbf{K}_3 = \begin{pmatrix} 1 \\ 1 \\ 0 \end{pmatrix}.$$

Then

$$\mathbf{X} = c_1 \begin{pmatrix} 1 \\ 1 \\ 1 \end{pmatrix} e^{t} + c_2 \begin{pmatrix} 1 \\ 0 \\ 1 \end{pmatrix} e^{2t} + c_3 \begin{pmatrix} 1 \\ 1 \\ 0 \end{pmatrix} e^{2t}.$$

24. We have $\det(\mathbf{A} - \lambda\mathbf{I}) = (1-\lambda)(\lambda - 2)^2 = 0$. For $\lambda_1 = 1$ we obtain

$$\mathbf{K}_1 = \begin{pmatrix} 1 \\ 0 \\ 0 \end{pmatrix}.$$

For $\lambda_2 = 2$ we obtain

$$\mathbf{K} = \begin{pmatrix} 0 \\ -1 \\ 1 \end{pmatrix}.$$

A solution of $(\mathbf{A} - \lambda_2\mathbf{I})\mathbf{P} = \mathbf{K}$ is

$$\mathbf{P} = \begin{pmatrix} 0 \\ -1 \\ 0 \end{pmatrix}$$

so that

$$\mathbf{X} = c_1 \begin{pmatrix} 1 \\ 0 \\ 0 \end{pmatrix} e^{t} + c_2 \begin{pmatrix} 0 \\ -1 \\ 1 \end{pmatrix} e^{2t} + c_3 \left[\begin{pmatrix} 0 \\ -1 \\ 1 \end{pmatrix} te^{2t} + \begin{pmatrix} 0 \\ -1 \\ 0 \end{pmatrix} e^{2t} \right].$$

27. We have $\det(\mathbf{A} - \lambda\mathbf{I}) = (\lambda - 4)^2 = 0$. For $\lambda_1 = 4$ we obtain

$$\mathbf{K} = \begin{pmatrix} 2 \\ 1 \end{pmatrix}.$$

A solution of $(\mathbf{A} - \lambda_1\mathbf{I})\mathbf{P} = \mathbf{K}$ is

$$\mathbf{P} = \begin{pmatrix} 1 \\ 1 \end{pmatrix}$$

so that

$$\mathbf{X} = c_1 \begin{pmatrix} 2 \\ 1 \end{pmatrix} e^{4t} + c_2 \left[\begin{pmatrix} 2 \\ 1 \end{pmatrix} te^{4t} + \begin{pmatrix} 1 \\ 1 \end{pmatrix} e^{4t} \right].$$

If

$$\mathbf{X}(0) = \begin{pmatrix} -1 \\ 6 \end{pmatrix}$$

then $c_1 = -7$ and $c_2 = 13$.

33. We have $\det(\mathbf{A} - \lambda\mathbf{I}) = \lambda^2 - 8\lambda + 17 = 0$. For $\lambda_1 = 4 + i$ we obtain

$$\mathbf{K}_1 = \begin{pmatrix} -1-i \\ 2 \end{pmatrix}$$

so that
$$\mathbf{X}_1 = \begin{pmatrix} -1-i \\ 2 \end{pmatrix} e^{(4+i)t} = \begin{pmatrix} \sin t - \cos t \\ 2\cos t \end{pmatrix} e^{4t} + i \begin{pmatrix} -\sin t - \cos t \\ 2\sin t \end{pmatrix} e^{4t}.$$

Then
$$\mathbf{X} = c_1 \begin{pmatrix} \sin t - \cos t \\ 2\cos t \end{pmatrix} e^{4t} + c_2 \begin{pmatrix} -\sin t - \cos t \\ 2\sin t \end{pmatrix} e^{4t}.$$

36. We have $\det(\mathbf{A} - \lambda\mathbf{I}) = \lambda^2 + 2\lambda + 5 = 0$. For $\lambda_1 = -1 + 2i$ we obtain

$$\mathbf{K}_1 = \begin{pmatrix} 2+2i \\ 1 \end{pmatrix}$$

so that
$$\mathbf{X}_1 = \begin{pmatrix} 2+2i \\ 1 \end{pmatrix} e^{(-1+2i)t}$$
$$= (\,2\cos 2t - 2\sin 2t \cos 2t\,)\, e^{-t} + i \begin{pmatrix} 2\cos 2t + 2\sin 2t \\ \sin 2t \end{pmatrix} e^{-t}.$$

Then
$$\mathbf{X} = c_1 \begin{pmatrix} 2\cos 2t - 2\sin 2t \\ \cos 2t \end{pmatrix} e^{-t} + c_2 \begin{pmatrix} 2\cos 2t + 2\sin 2t \\ \sin 2t \end{pmatrix} e^{-t}.$$

39. We have $\det(\mathbf{A} - \lambda\mathbf{I}) = (1-\lambda)(\lambda^2 - 2\lambda + 2) = 0$. For $\lambda_1 = 1$ we obtain

$$\mathbf{K}_1 = \begin{pmatrix} 0 \\ 2 \\ 1 \end{pmatrix}.$$

For $\lambda_2 = 1 + i$ we obtain

$$\mathbf{K}_2 = \begin{pmatrix} 1 \\ i \\ i \end{pmatrix}$$

so that
$$\mathbf{X}_2 = \begin{pmatrix} 1 \\ i \\ i \end{pmatrix} e^{(1+i)t} = \begin{pmatrix} \cos t \\ -\sin t \\ -\sin t \end{pmatrix} e^{t} + i \begin{pmatrix} \sin t \\ \cos t \\ \cos t \end{pmatrix} e^{t}.$$

Then
$$\mathbf{X} = c_1 \begin{pmatrix} 0 \\ 2 \\ 1 \end{pmatrix} e^{t} + c_2 \begin{pmatrix} \cos t \\ -\sin t \\ -\sin t \end{pmatrix} e^{t} + c_3 \begin{pmatrix} \sin t \\ \cos t \\ \cos t \end{pmatrix} e^{t}.$$

42. We have $\det(\mathbf{A}-\lambda\mathbf{I}) = -(\lambda+2)(\lambda^2+4) = 0$. For $\lambda_1 = -2$ we obtain

$$\mathbf{K}_1 = \begin{pmatrix} 0 \\ -1 \\ 1 \end{pmatrix}.$$

For $\lambda_2 = 2i$ we obtain

$$\mathbf{K}_2 = \begin{pmatrix} -2-2i \\ 1 \\ 1 \end{pmatrix}$$

so that $$\mathbf{X}_2 = \begin{pmatrix} -2-2i \\ 1 \\ 1 \end{pmatrix} e^{2it} = \begin{pmatrix} -2\cos 2t + 2\sin 2t \\ \cos 2t \\ \cos 2t \end{pmatrix} + i \begin{pmatrix} -2\cos 2t - 2\sin 2t \\ \sin 2t \\ \sin 2t \end{pmatrix}.$$

Then $$\mathbf{X} = c_1 \begin{pmatrix} 0 \\ -1 \\ 1 \end{pmatrix} e^{-2t} + c_2 \begin{pmatrix} -2\cos 2t + 2\sin 2t \\ \cos 2t \\ \cos 2t \end{pmatrix} + c_3 \begin{pmatrix} -2\cos 2t - 2\sin 2t \\ \sin 2t \\ \sin 2t \end{pmatrix}.$$

Exercises 8.3

3. From

$$\mathbf{X}' = \begin{pmatrix} 3 & -5 \\ 3/4 & -1 \end{pmatrix}\mathbf{X} + \begin{pmatrix} 1 \\ -1 \end{pmatrix} e^{t/2}$$

we obtain $$\mathbf{X}_c = c_1 \begin{pmatrix} 10 \\ 3 \end{pmatrix} e^{3t/2} + c_2 \begin{pmatrix} 2 \\ 1 \end{pmatrix} e^{t/2}.$$

Then $$\mathbf{\Phi} = \begin{pmatrix} 10e^{3t/2} & 2e^{t/2} \\ 3e^{3t/2} & e^{t/2} \end{pmatrix} \quad \text{and} \quad \mathbf{\Phi}^{-1} = \begin{pmatrix} \frac{1}{4}e^{-3t/2} & -\frac{1}{2}e^{-3t/2} \\ -\frac{3}{4}e^{-t/2} & \frac{5}{2}e^{-t/2} \end{pmatrix}$$

so that $$\mathbf{U} = \int \mathbf{\Phi}^{-1}\mathbf{F}\,dt = \int \begin{pmatrix} \frac{3}{4}e^{-t} \\ -\frac{13}{4} \end{pmatrix} dt = \begin{pmatrix} -\frac{3}{4}e^{-t} \\ -\frac{13}{4}t \end{pmatrix}$$

and $$\mathbf{X}_p = \mathbf{\Phi}\mathbf{U} = \begin{pmatrix} -13/2 \\ -13/4 \end{pmatrix} te^{t/2} + \begin{pmatrix} -15/2 \\ -9/4 \end{pmatrix} e^{t/2}.$$

6. From

$$\mathbf{X}' = \begin{pmatrix} 0 & 2 \\ -1 & 3 \end{pmatrix}\mathbf{X} + \begin{pmatrix} 2 \\ e^{-3t} \end{pmatrix}$$

we obtain $$\mathbf{X}_c = c_1 \begin{pmatrix} 2 \\ 1 \end{pmatrix} e^{t} + c_2 \begin{pmatrix} 1 \\ 1 \end{pmatrix} e^{2t}.$$

Then
$$\mathbf{\Phi} = \begin{pmatrix} 2e^t & e^{2t} \\ e^t & e^{2t} \end{pmatrix} \quad \text{and} \quad \mathbf{\Phi}^{-1} = \begin{pmatrix} e^{-t} & -e^{-t} \\ -e^{-2t} & 2e^{-2t} \end{pmatrix}$$

so that
$$\mathbf{U} = \int \mathbf{\Phi}^{-1}\mathbf{F}\,dt = \int \begin{pmatrix} 2e^{-t} - e^{-4t} \\ -2e^{-2t} + 2e^{-5t} \end{pmatrix} dt = \begin{pmatrix} -2e^{-t} + \frac{1}{4}e^{-4t} \\ e^{-2t} - \frac{2}{5}e^{-5t} \end{pmatrix}$$

and
$$\mathbf{X}_p = \mathbf{\Phi}\mathbf{U} = \begin{pmatrix} \frac{1}{10}e^{-3t} - 3 \\ -\frac{3}{20}e^{-3t} - 1 \end{pmatrix}.$$

9. From
$$\mathbf{X}' = \begin{pmatrix} 3 & 2 \\ -2 & -1 \end{pmatrix}\mathbf{X} + \begin{pmatrix} 2 \\ 1 \end{pmatrix} e^{-t}$$

we obtain
$$\mathbf{X}_c = c_1 \begin{pmatrix} 1 \\ -1 \end{pmatrix} e^t + c_2 \left[\begin{pmatrix} 1 \\ -1 \end{pmatrix} te^t + \begin{pmatrix} 0 \\ 1/2 \end{pmatrix} e^t \right].$$

Then
$$\mathbf{\Phi} = \begin{pmatrix} e^t & te^t \\ -e^t & \frac{1}{2}e^t - te^t \end{pmatrix} \quad \text{and} \quad \mathbf{\Phi}^{-1} = \begin{pmatrix} e^{-t} - 2te^{-t} & -2te^{-t} \\ 2e^{-t} & 2e^{-t} \end{pmatrix}$$

so that
$$\mathbf{U} = \int \mathbf{\Phi}^{-1}\mathbf{F}\,dt = \int \begin{pmatrix} 2e^{-2t} - 6te^{-2t} \\ 6e^{-2t} \end{pmatrix} dt = \begin{pmatrix} \frac{1}{2}e^{-2t} + 3te^{-2t} \\ -3e^{-2t} \end{pmatrix}$$

and
$$\mathbf{X}_p = \mathbf{\Phi}\mathbf{U} = \begin{pmatrix} 1/2 \\ -2 \end{pmatrix} e^{-t}.$$

12. From
$$\mathbf{X}' = \begin{pmatrix} 1 & -1 \\ 1 & 1 \end{pmatrix}\mathbf{X} + \begin{pmatrix} 3 \\ 3 \end{pmatrix} e^t$$

we obtain
$$\mathbf{X}_c = c_1 \begin{pmatrix} -\sin t \\ \cos t \end{pmatrix} e^t + c_2 \begin{pmatrix} \cos t \\ \sin t \end{pmatrix} e^t.$$

Then
$$\mathbf{\Phi} = \begin{pmatrix} -\sin t & \cos t \\ \cos t & \sin t \end{pmatrix} e^t \quad \text{and} \quad \mathbf{\Phi}^{-1} = \begin{pmatrix} -\sin t & \cos t \\ \cos t & \sin t \end{pmatrix} e^{-t}$$

so that
$$\mathbf{U} = \int \mathbf{\Phi}^{-1}\mathbf{F}\,dt = \int \begin{pmatrix} -3\sin t + 3\cos t \\ 3\cos t + 3\sin t \end{pmatrix} dt = \begin{pmatrix} 3\cos t + 3\sin t \\ 3\sin t - 3\cos t \end{pmatrix}$$

and
$$\mathbf{X}_p = \mathbf{\Phi}\mathbf{U} = \begin{pmatrix} -3 \\ 3 \end{pmatrix} e^t.$$

15. From
$$\mathbf{X}' = \begin{pmatrix} 0 & 1 \\ -1 & 0 \end{pmatrix}\mathbf{X} + \begin{pmatrix} 0 \\ \sec t \tan t \end{pmatrix}$$

we obtain $$\mathbf{X}_c = c_1\begin{pmatrix}\cos t\\ -\sin t\end{pmatrix} + c_2\begin{pmatrix}\sin t\\ \cos t\end{pmatrix}.$$

Then $$\mathbf{\Phi} = \begin{pmatrix}\cos t & \sin t\\ -\sin t & \cos t\end{pmatrix}t \quad\text{and}\quad \mathbf{\Phi}^{-1} = \begin{pmatrix}\cos t & -\sin t\\ \sin t & \cos t\end{pmatrix}$$

so that $$\mathbf{U} = \int \mathbf{\Phi}^{-1}\mathbf{F}\,dt = \int\begin{pmatrix}-\tan^2 t\\ \tan t\end{pmatrix}dt = \begin{pmatrix}t-\tan t\\ \ln|\sec t|\end{pmatrix}$$

and $$\mathbf{X}_p = \mathbf{\Phi}\mathbf{U} = \begin{pmatrix}\cos t\\ -\sin t\end{pmatrix}t + \begin{pmatrix}-\sin t\\ \sin t\tan t\end{pmatrix} + \begin{pmatrix}\sin t\\ \cos t\end{pmatrix}\ln|\sec t|.$$

18. From
$$\mathbf{X}' = \begin{pmatrix}1 & -2\\ 1 & -1\end{pmatrix}\mathbf{X} + \begin{pmatrix}\tan t\\ 1\end{pmatrix}$$

we obtain $$\mathbf{X}_c = c_1\begin{pmatrix}\cos t - \sin t\\ \cos t\end{pmatrix} + c_2\begin{pmatrix}\cos t + \sin t\\ \sin t\end{pmatrix}.$$

Then $$\mathbf{\Phi} = \begin{pmatrix}\cos t - \sin t & \cos t + \sin t\\ \cos t & \sin t\end{pmatrix} \quad\text{and}\quad \mathbf{\Phi}^{-1} = \begin{pmatrix}-\sin t & \cos t + \sin t\\ \cos t & \sin t - \cos t\end{pmatrix}$$

so that $$\mathbf{U} = \int \mathbf{\Phi}^{-1}\mathbf{F}\,dt = \int\begin{pmatrix}2\cos t + \sin t - \sec t\\ 2\sin t - \cos t\end{pmatrix}dt = \begin{pmatrix}2\sin t - \cos t - \ln|\sec t + \tan t|\\ -2\cos t - \sin t\end{pmatrix}$$

and $$\mathbf{X}_p = \mathbf{\Phi}\mathbf{U} = \begin{pmatrix}3\sin t\cos t - \cos^2 t - 2\sin^2 t + (\sin t - \cos t)\ln|\sec t + \tan t|\\ \sin^2 t - \cos^2 t - \cos t(\ln|\sec t + \tan t|)\end{pmatrix}.$$

21. From
$$\mathbf{X}' = \begin{pmatrix}3 & -1\\ -1 & 3\end{pmatrix}\mathbf{X} + \begin{pmatrix}4e^{2t}\\ 4e^{4t}\end{pmatrix}$$

we obtain $$\mathbf{\Phi} = \begin{pmatrix}-e^{4t} & e^{2t}\\ e^{4t} & e^{2t}\end{pmatrix},\quad \mathbf{\Phi}^{-1} = \begin{pmatrix}-\frac{1}{2}e^{-4t} & \frac{1}{2}e^{-4t}\\ \frac{1}{2}e^{-2t} & \frac{1}{2}e^{-2t}\end{pmatrix},$$

and $$\mathbf{X} = \mathbf{\Phi}\mathbf{\Phi}^{-1}(0)\mathbf{X}(0) + \mathbf{\Phi}\int_0^t \mathbf{\Phi}^{-1}\mathbf{F}\,ds = \mathbf{\Phi}\cdot\begin{pmatrix}0\\ 1\end{pmatrix} + \mathbf{\Phi}\cdot\begin{pmatrix}e^{-2t}+2t-1\\ e^{2t}+2t-1\end{pmatrix}$$
$$= \begin{pmatrix}2\\ 2\end{pmatrix}te^{2t} + \begin{pmatrix}-1\\ 1\end{pmatrix}e^{2t} + \begin{pmatrix}-2\\ 2\end{pmatrix}te^{4t} + \begin{pmatrix}2\\ 0\end{pmatrix}e^{4t}.$$

24. **(a)** The eigenvalues are 0, 1, 3, and 4, with corresponding eigenvectors

$$\begin{pmatrix} -6 \\ -4 \\ 1 \\ 2 \end{pmatrix}, \quad \begin{pmatrix} 2 \\ 1 \\ 0 \\ 0 \end{pmatrix}, \quad \begin{pmatrix} 3 \\ 1 \\ 2 \\ 1 \end{pmatrix}, \quad \text{and} \quad \begin{pmatrix} -1 \\ 1 \\ 0 \\ 0 \end{pmatrix}.$$

(b) $$\mathbf{\Phi} = \begin{pmatrix} -6 & 2e^t & 3e^{3t} & -e^{4t} \\ -4 & e^t & e^{3t} & e^{4t} \\ 1 & 0 & 2e^{3t} & 0 \\ 2 & 0 & e^{3t} & 0 \end{pmatrix}, \qquad \mathbf{\Phi}^{-1} = \begin{pmatrix} 0 & 0 & -\frac{1}{3} & \frac{2}{3} \\ \frac{1}{3}e^{-t} & \frac{1}{3}e^{-t} & -2e^{-t} & \frac{8}{3}e^{-t} \\ 0 & 0 & \frac{2}{3}e^{-3t} & -\frac{1}{3}e^{-3t} \\ -\frac{1}{3}e^{-4t} & \frac{2}{3}e^{-4t} & 0 & \frac{1}{3}e^{-4t} \end{pmatrix}$$

(c) $$\mathbf{\Phi}^{-1}(t)\mathbf{F}(t) = \begin{pmatrix} \frac{2}{3} - \frac{1}{3}e^{2t} \\ \frac{1}{3}e^{-2t} + \frac{8}{3}e^{-t} - 2e^t + \frac{1}{3}t \\ -\frac{1}{3}e^{-3t} + \frac{2}{3}e^{-t} \\ \frac{2}{3}e^{-5t} + \frac{1}{3}e^{-4t} - \frac{1}{3}te^{-3t} \end{pmatrix},$$

$$\int \mathbf{\Phi}^{-1}(t)\mathbf{F}(t)dt = \begin{pmatrix} -\frac{1}{6}e^{2t} + \frac{2}{3}t \\ -\frac{1}{6}e^{-2t} - \frac{8}{3}e^{-t} - 2e^t + \frac{1}{6}t^2 \\ \frac{1}{9}e^{-3t} - \frac{2}{3}e^{-t} \\ -\frac{2}{15}e^{-5t} - \frac{1}{12}e^{-4t} + \frac{1}{27}e^{-3t} + \frac{1}{9}te^{-3t} \end{pmatrix},$$

$$\mathbf{\Phi}(t)\int \mathbf{\Phi}^{-1}(t)\mathbf{F}(t)dt = \begin{pmatrix} -5e^{2t} - \frac{1}{5}e^{-t} - \frac{1}{27}e^t - \frac{1}{9}te^t + \frac{1}{3}t^2e^t - 4t - \frac{59}{12} \\ -2e^{2t} - \frac{3}{10}e^{-t} + \frac{1}{27}e^t + \frac{1}{9}te^t + \frac{1}{6}t^2e^t - \frac{8}{3}t - \frac{95}{36} \\ -\frac{3}{2}e^{2t} + \frac{2}{3}t + \frac{2}{9} \\ -e^{2t} + \frac{4}{3}t - \frac{1}{9} \end{pmatrix},$$

$$\mathbf{\Phi}(t)\mathbf{C} = \begin{pmatrix} -6c_1 + 2c_2e^t + 3c_3e^{3t} - c_4e^{4t} \\ -4c_1 + c_2e^t + c_3e^{3t} + c_4e^{4t} \\ c_1 + 2c_3e^{3t} \\ 2c_1 + c_3e^{3t} \end{pmatrix},$$

$$\mathbf{\Phi}(t)\mathbf{C} + \mathbf{\Phi}(t)\int \mathbf{\Phi}^{-1}(t)\mathbf{F}(t)dt$$

$$= \begin{pmatrix} -6c_1 + 2c_2e^t + 3c_3e^{3t} - c_4e^{4t} \\ -4c_1 + c_2e^t + c_3e^{3t} + c_4e^{4t} \\ c_1 + 2c_3e^{3t} \\ 2c_1 + c_3e^{3t} \end{pmatrix} + \begin{pmatrix} -5e^{2t} - \frac{1}{5}e^{-t} - \frac{1}{27}e^t - \frac{1}{9}te^t + \frac{1}{3}t^2e^t - 4t - \frac{59}{12} \\ -2e^{2t} - \frac{3}{10}e^{-t} + \frac{1}{27}e^t + \frac{1}{9}te^t + \frac{1}{6}t^2e^t - \frac{8}{3}t - \frac{95}{36} \\ -\frac{3}{2}e^{2t} + \frac{2}{3}t + \frac{2}{9} \\ -e^{2t} + \frac{4}{3}t - \frac{1}{9} \end{pmatrix}$$

(d) $$\mathbf{X}(t) = c_1\begin{pmatrix}-6\\-4\\1\\2\end{pmatrix} + c_2\begin{pmatrix}2\\1\\0\\0\end{pmatrix} + c_3\begin{pmatrix}3\\1\\2\\1\end{pmatrix} + c_4\begin{pmatrix}-1\\1\\0\\0\end{pmatrix}$$

$$+\begin{pmatrix}-5e^{2t} - \frac{1}{5}e^{-t} - \frac{1}{27}e^t - \frac{1}{9}te^t + \frac{1}{3}t^2e^t - 4t - \frac{59}{12}\\ -2e^{2t} - \frac{3}{10}e^{-t} + \frac{1}{27}e^t + \frac{1}{9}te^t + \frac{1}{6}t^2e^t - \frac{8}{3}t - \frac{95}{36}\\ -\frac{3}{2}e^{2t} + \frac{2}{3}t + \frac{2}{9}\\ -e^{2t} + \frac{4}{3}t - \frac{1}{9}\end{pmatrix}$$

Exercises 8.4

3. For

$$\mathbf{A} = \begin{pmatrix}1 & 1 & 1\\ 1 & 1 & 1\\ -2 & -2 & -2\end{pmatrix}$$

we have

$$\mathbf{A}^2 = \begin{pmatrix}1 & 1 & 1\\ 1 & 1 & 1\\ -2 & -2 & -2\end{pmatrix}\begin{pmatrix}1 & 1 & 1\\ 1 & 1 & 1\\ -2 & -2 & -2\end{pmatrix} = \begin{pmatrix}0 & 0 & 0\\ 0 & 0 & 0\\ 0 & 0 & 0\end{pmatrix}.$$

Thus, $\mathbf{A}^3 = \mathbf{A}^4 = \mathbf{A}^5 = \cdots = \mathbf{0}$ and

$$e^{\mathbf{A}t} = \mathbf{I} + \mathbf{A}t = \begin{pmatrix}1 & 0 & 0\\ 0 & 1 & 0\\ 0 & 0 & 1\end{pmatrix} + \begin{pmatrix}t & t & t\\ t & t & t\\ -2t & -2t & -2t\end{pmatrix} = \begin{pmatrix}t+1 & t & t\\ t & t+1 & t\\ -2t & -2t & -2t+1\end{pmatrix}.$$

6. For $\mathbf{A} = \begin{pmatrix}0 & 1\\ 1 & 0\end{pmatrix}$ we have

$$e^{t\mathbf{A}} = \begin{pmatrix}\cosh t & \sinh t\\ \sinh t & \cosh t\end{pmatrix}.$$

Then

$$\mathbf{X} = \begin{pmatrix}\cosh t & \sinh t\\ \sinh t & \cosh t\end{pmatrix}\begin{pmatrix}c_1\\ c_2\end{pmatrix} = c_1\begin{pmatrix}\cosh t\\ \sinh t\end{pmatrix} + c_2\begin{pmatrix}\sinh t\\ \cosh t\end{pmatrix}.$$

9. To solve

$$\mathbf{X}' = \begin{pmatrix}1 & 0\\ 0 & 2\end{pmatrix}\mathbf{X} + \begin{pmatrix}3\\ -1\end{pmatrix}$$

we identify $t_0 = 0$, $\mathbf{F}(s) = \begin{pmatrix}3\\ -1\end{pmatrix}$, and use the results of Problem 1 and equation (5) in the text.

$$\mathbf{X}(t) = e^{\mathbf{A}t}\mathbf{C} + e^{\mathbf{A}t}\int_{t_0}^{t} e^{-\mathbf{A}s}\mathbf{F}(s)\,ds$$

$$= \begin{pmatrix} e^t & 0 \\ 0 & e^{2t} \end{pmatrix}\begin{pmatrix} c_1 \\ c_2 \end{pmatrix} + \begin{pmatrix} e^t & 0 \\ 0 & e^{2t} \end{pmatrix}\int_0^t \begin{pmatrix} e^{-s} & 0 \\ 0 & e^{-2s} \end{pmatrix}\begin{pmatrix} 3 \\ -1 \end{pmatrix} ds$$

$$= \begin{pmatrix} c_1e^t \\ c_2e^{2t} \end{pmatrix} + \begin{pmatrix} e^t & 0 \\ 0 & e^{2t} \end{pmatrix}\int_0^t \begin{pmatrix} 3e^{-s} \\ -e^{-2s} \end{pmatrix} ds$$

$$= \begin{pmatrix} c_1e^t \\ c_2e^{2t} \end{pmatrix} + \begin{pmatrix} e^t & 0 \\ 0 & e^{2t} \end{pmatrix}\begin{pmatrix} -3e^{-s} \\ \frac{1}{2}e^{-2s} \end{pmatrix}\Bigg|_0^t$$

$$= \begin{pmatrix} c_1e^t \\ c_2e^{2t} \end{pmatrix} + \begin{pmatrix} e^t & 0 \\ 0 & e^{2t} \end{pmatrix}\begin{pmatrix} -3e^{-t} - 3 \\ \frac{1}{2}e^{-2t} - \frac{1}{2} \end{pmatrix}$$

$$= \begin{pmatrix} c_1e^t \\ c_2e^{2t} \end{pmatrix} + \begin{pmatrix} -3 - 3e^t \\ \frac{1}{2} - \frac{1}{2}e^{2t} \end{pmatrix} = c_3\begin{pmatrix} 1 \\ 0 \end{pmatrix}e^t + c_4\begin{pmatrix} 0 \\ 1 \end{pmatrix}e^{2t} + \begin{pmatrix} -3 \\ \frac{1}{2} \end{pmatrix}.$$

12. To solve

$$\mathbf{X}' = \begin{pmatrix} 0 & 1 \\ 1 & 0 \end{pmatrix}\mathbf{X} + \begin{pmatrix} \cosh t \\ \sinh t \end{pmatrix}$$

we identify $t_0 = 0$, $\mathbf{F}(s) = \begin{pmatrix} \cosh t \\ \sinh t \end{pmatrix}$, and use equation (5) in the text and the value of $e^{t\mathbf{A}}$ given in the solution of Problem 6.

$$\mathbf{X}(t) = e^{\mathbf{A}t}\mathbf{C} + e^{\mathbf{A}t}\int_{t_0}^{t} e^{-\mathbf{A}s}\mathbf{F}(s)\,ds$$

$$= \begin{pmatrix} \cosh t & \sinh t \\ \sinh t & \cosh t \end{pmatrix}\begin{pmatrix} c_1 \\ c_2 \end{pmatrix} + \begin{pmatrix} \cosh t & \sinh t \\ \sinh t & \cosh t \end{pmatrix}\int_0^t \begin{pmatrix} \cosh s & -\sinh s \\ -\sinh s & \cosh s \end{pmatrix}\begin{pmatrix} \cosh s \\ \sinh s \end{pmatrix} ds$$

$$= \begin{pmatrix} c_1\cosh t + c_2\sinh t \\ c_1\sinh t + c_2\cosh t \end{pmatrix} + \begin{pmatrix} \cosh t & \sinh t \\ \sinh t & \cosh t \end{pmatrix}\int_0^t \begin{pmatrix} 1 \\ 0 \end{pmatrix} ds$$

$$= \begin{pmatrix} c_1\cosh t + c_2\sinh t \\ c_1\sinh t + c_2\cosh t \end{pmatrix} + \begin{pmatrix} \cosh t & \sinh t \\ \sinh t & \cosh t \end{pmatrix}\begin{pmatrix} s \\ 0 \end{pmatrix}\Bigg|_0^t$$

$$= \begin{pmatrix} c_1\cosh t + c_2\sinh t \\ c_1\sinh t + c_2\cosh t \end{pmatrix} + \begin{pmatrix} \cosh t & \sinh t \\ \sinh t & \cosh t \end{pmatrix}\begin{pmatrix} t \\ 0 \end{pmatrix}$$

$$= \begin{pmatrix} c_1\cosh t + c_2\sinh t \\ c_1\sinh t + c_2\cosh t \end{pmatrix} + \begin{pmatrix} t\cosh t \\ t\sinh t \end{pmatrix} = c_1\begin{pmatrix} \cosh t \\ \sinh t \end{pmatrix} + c_2\begin{pmatrix} \sinh t \\ \cosh t \end{pmatrix} + t\begin{pmatrix} \cosh t \\ \sinh t \end{pmatrix}.$$

15. Solving

$$\begin{vmatrix} 2-\lambda & 1 \\ -3 & 6-\lambda \end{vmatrix} = \lambda^2 - 8\lambda + 15 = (\lambda-3)(\lambda-5) = 0$$

we find eigenvalues $\lambda_1 = 3$ and $\lambda_2 = 5$. Corresponding eigenvectors are

$$\mathbf{K}_1 = \begin{pmatrix} 1 \\ 1 \end{pmatrix} \quad \text{and} \quad \mathbf{K}_2 = \begin{pmatrix} 1 \\ 3 \end{pmatrix}.$$

Then

$$\mathbf{P} = \begin{pmatrix} 1 & 1 \\ 1 & 3 \end{pmatrix}, \quad \mathbf{P}^{-1} = \begin{pmatrix} 3/2 & -1/2 \\ -1/2 & 1/2 \end{pmatrix}, \quad \text{and} \quad \mathbf{D} = \begin{pmatrix} 3 & 0 \\ 0 & 5 \end{pmatrix},$$

so

$$\mathbf{PDP}^{-1} = \begin{pmatrix} 2 & 1 \\ -3 & 6 \end{pmatrix}.$$

18. From equation (3) in the text

$$e^{\mathbf{D}t} = \begin{pmatrix} 1 & 0 & \cdots & 0 \\ 0 & 1 & \cdots & 0 \\ \vdots & \vdots & \ddots & \vdots \\ 0 & 0 & \cdots & 1 \end{pmatrix} + \begin{pmatrix} \lambda_1 & 0 & \cdots & 0 \\ 0 & \lambda_2 & \cdots & 0 \\ \vdots & \vdots & \ddots & \vdots \\ 0 & 0 & \cdots & \lambda_n \end{pmatrix} + \frac{1}{2!}t^2 \begin{pmatrix} \lambda_1^2 & 0 & \cdots & 0 \\ 0 & \lambda_2^2 & \cdots & 0 \\ \vdots & \vdots & \ddots & \vdots \\ 0 & 0 & \cdots & \lambda_n^2 \end{pmatrix}$$

$$+ \frac{1}{3!}t^3 \begin{pmatrix} \lambda_1^3 & 0 & \cdots & 0 \\ 0 & \lambda_2^3 & \cdots & 0 \\ \vdots & \vdots & \ddots & \vdots \\ 0 & 0 & \cdots & \lambda_n^3 \end{pmatrix} + \cdots$$

$$= \begin{pmatrix} 1 + \lambda_1 t + \frac{1}{2!}(\lambda_1 t)^2 + \cdots & 0 & \cdots & 0 \\ 0 & 1 + \lambda_2 t + \frac{1}{2!}(\lambda_2 t)^2 + \cdots & \cdots & 0 \\ \vdots & \vdots & \ddots & \vdots \\ 0 & 0 & \cdots & 1 + \lambda_n t + \frac{1}{2!}(\lambda_n t)^2 + \cdots \end{pmatrix}$$

$$= \begin{pmatrix} e^{\lambda_1 t} & 0 & \cdots & 0 \\ 0 & e^{\lambda_2 t} & \cdots & 0 \\ \vdots & \vdots & \ddots & \vdots \\ 0 & 0 & \cdots & e^{\lambda_n t} \end{pmatrix}.$$

Chapter 8 Review Exercises

3. We have $\det(\mathbf{A}-\lambda\mathbf{I}) = (\lambda-1)^2 = 0$ and $\mathbf{K} = \begin{pmatrix} 1 \\ -1 \end{pmatrix}$. A solution to $(\mathbf{A}-\lambda\mathbf{I})\mathbf{P} = \mathbf{K}$ is $\mathbf{P} = \begin{pmatrix} 0 \\ 1 \end{pmatrix}$ so that

$$\mathbf{X} = c_1 \begin{pmatrix} 1 \\ -1 \end{pmatrix} e^t + c_2 \left[\begin{pmatrix} 1 \\ -1 \end{pmatrix} te^t + \begin{pmatrix} 0 \\ 1 \end{pmatrix} e^t \right].$$

6. We have $\det(\mathbf{A}-\lambda\mathbf{I}) = \lambda^2 - 2\lambda + 2 = 0$. For $\lambda = 1+i$ we obtain $\mathbf{K}_1 = \begin{pmatrix} 3-i \\ 2 \end{pmatrix}$ and

$$\mathbf{X}_1 = \begin{pmatrix} 3-i \\ 2 \end{pmatrix} e^{(1+i)t} = \begin{pmatrix} 3\cos t + \sin t \\ 2\cos t \end{pmatrix} e^t + i \begin{pmatrix} -\cos t + 3\sin t \\ 2\sin t \end{pmatrix} e^t.$$

Then

$$\mathbf{X} = c_1 \begin{pmatrix} 3\cos t + \sin t \\ 2\cos t \end{pmatrix} e^t + c_2 \begin{pmatrix} -\cos t + 3\sin t \\ 2\sin t \end{pmatrix} e^t.$$

9. We have

$$\mathbf{X}_c = c_1 \begin{pmatrix} 1 \\ 0 \end{pmatrix} e^{2t} + c_2 \begin{pmatrix} 4 \\ 1 \end{pmatrix} e^{4t}.$$

Then

$$\mathbf{\Phi} = \begin{pmatrix} e^{2t} & 4e^{4t} \\ 0 & e^{4t} \end{pmatrix}, \quad \mathbf{\Phi}^{-1} = \begin{pmatrix} e^{-2t} & -4e^{-2t} \\ 0 & e^{-4t} \end{pmatrix},$$

and

$$\mathbf{U} = \int \mathbf{\Phi}^{-1}\mathbf{F}\,dt = \int \begin{pmatrix} 2e^{-2t} - 64te^{-2t} \\ 16te^{-4t} \end{pmatrix} dt = \begin{pmatrix} 15e^{-2t} + 32te^{-2t} \\ -e^{-4t} - 4te^{-4t} \end{pmatrix},$$

so that

$$\mathbf{X}_p = \mathbf{\Phi}\mathbf{U} = \begin{pmatrix} 11 + 16t \\ -1 - 4t \end{pmatrix}.$$

12. We have

$$\mathbf{X}_c = c_1 \begin{pmatrix} 1 \\ -1 \end{pmatrix} e^{2t} + c_2 \left[\begin{pmatrix} 1 \\ -1 \end{pmatrix} te^{2t} + \begin{pmatrix} 1 \\ 0 \end{pmatrix} e^{2t} \right].$$

Then

$$\mathbf{\Phi} = \begin{pmatrix} e^{2t} & te^{2t} + e^{2t} \\ -e^{2t} & -te^{2t} \end{pmatrix}, \quad \mathbf{\Phi}^{-1} = \begin{pmatrix} -te^{-2t} & -te^{-2t} - e^{-2t} \\ e^{-2t} & e^{-2t} \end{pmatrix},$$

and

$$\mathbf{U} = \int \mathbf{\Phi}^{-1}\mathbf{F}\,dt = \int \begin{pmatrix} t-1 \\ -1 \end{pmatrix} dt = \begin{pmatrix} \frac{1}{2}t^2 - t \\ -t \end{pmatrix},$$

so that

$$\mathbf{X}_p = \mathbf{\Phi}\mathbf{U} = \begin{pmatrix} -1/2 \\ 1/2 \end{pmatrix} t^2 e^{2t} + \begin{pmatrix} -2 \\ 1 \end{pmatrix} te^{2t}.$$

9 Numerical Methods for Ordinary Differential Equations

Exercises 9.1

3.

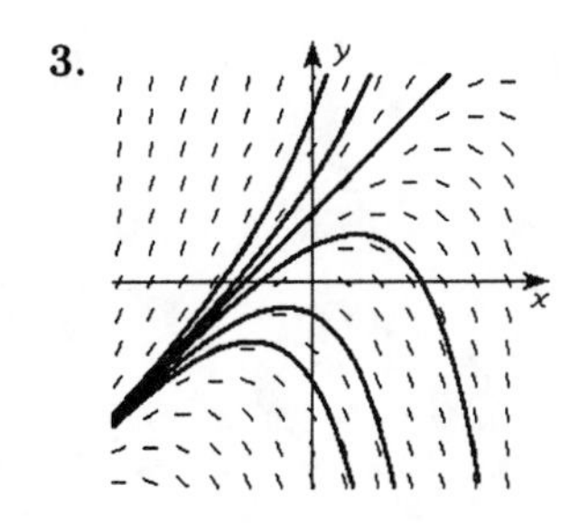

6. Setting $x + y = c$ we obtain the isoclines $y = -x + c$.

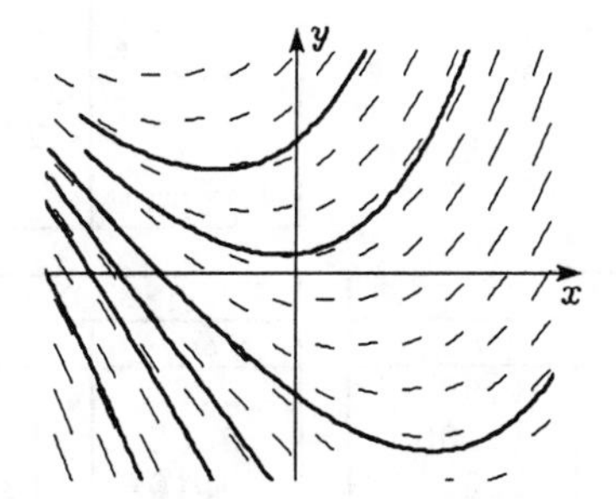

9. Setting $0.2x^2 + y = c$ we obtain the isoclines $y = c - 0.2x^2$.

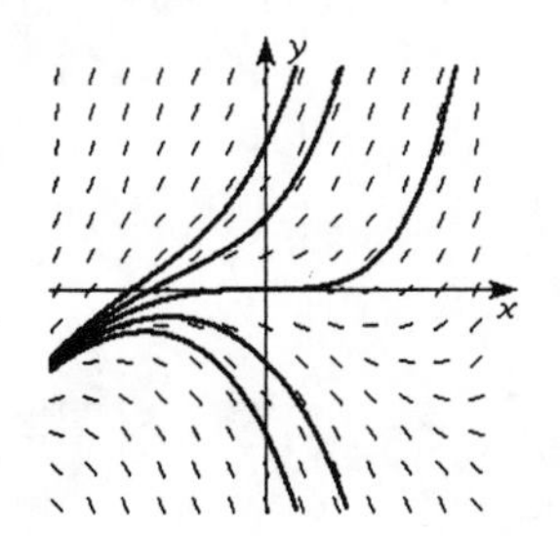

12. Setting $1 - y/x = c$ we obtain the isoclines $y = (1 - c)x$.

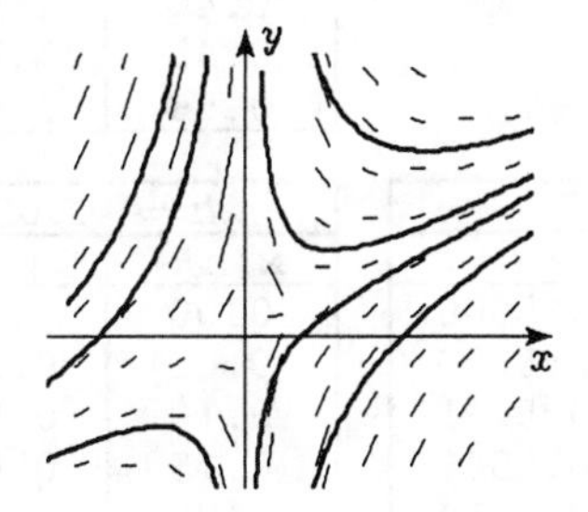

Exercises 9.2

All tables in this chapter were constructed in a spreadsheet program which does not support subscripts. Consequently, x_n and y_n will be indicated as $x(n)$ and $y(n)$, respectively.

3.

h = 0.1	
x(n)	y(n)
1.00	5.0000
1.10	3.8000
1.20	2.9800
1.30	2.4260
1.40	2.0582
1.50	1.8207

h = 0.05	
x(n)	y(n)
1.00	5.0000
1.05	4.4000
1.10	3.8950
1.15	3.4708
1.20	3.1151
1.25	2.8179
1.30	2.5702
1.35	2.3647
1.40	2.1950
1.45	2.0557
1.50	1.9424

6.

h = 0.1	
x(n)	y(n)
0.00	1.0000
0.10	1.1000
0.20	1.2220
0.30	1.3753
0.40	1.5735
0.50	1.8371

h = 0.05	
x(n)	y(n)
0.00	1.0000
0.05	1.0500
0.10	1.1053
0.15	1.1668
0.20	1.2360
0.25	1.3144
0.30	1.4039
0.35	1.5070
0.40	1.6267
0.45	1.7670
0.50	1.9332

9.

h = 0.1	
x(n)	y(n)
0.00	0.5000
0.10	0.5250
0.20	0.5431
0.30	0.5548
0.40	0.5613
0.50	0.5639

h = 0.05	
x(n)	y(n)
0.00	0.5000
0.05	0.5125
0.10	0.5232
0.15	0.5322
0.20	0.5395
0.25	0.5452
0.30	0.5496
0.35	0.5527
0.40	0.5547
0.45	0.5559
0.50	0.5565

12.

$h =$ 0.1	
x(n)	y(n)
0.00	0.5000
0.10	0.5250
0.20	0.5499
0.30	0.5747
0.40	0.5991
0.50	0.6231

$h =$ 0.05	
x(n)	y(n)
0.00	0.5000
0.05	0.5125
0.10	0.5250
0.15	0.5375
0.20	0.5499
0.25	0.5623
0.30	0.5746
0.35	0.5868
0.40	0.5989
0.45	0.6109
0.50	0.6228

15. (a)

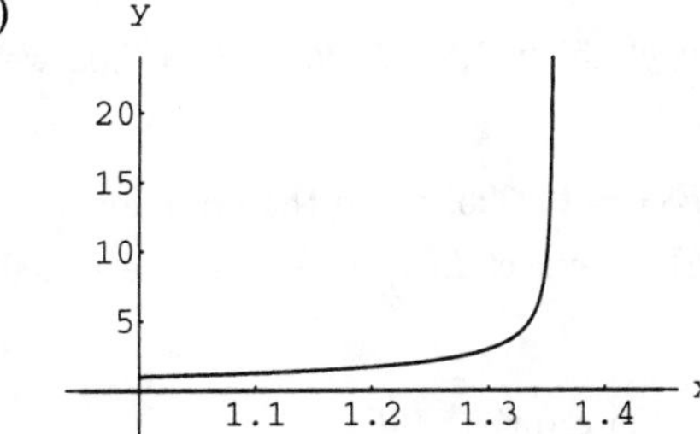

(b)

h=0.1	EULER	IMPROVED EULER
x(n)	y(n)	y(n)
1.00	1.0000	1.0000
1.10	1.2000	1.2469
1.20	1.4938	1.6668
1.30	1.9711	2.6427
1.40	2.9060	8.7988

18. (a) Using the improved Euler method we obtain $y(0.1) \approx y_1 = 1.22$.

(b) Using $y''' = 8e^{2x}$ we see that the local truncation error is

$$y'''(c)\,\frac{h^3}{6} = 8e^{2c}\,\frac{(0.1)^3}{6} = 0.001333e^{2c}.$$

Since e^{2x} is an increasing function, $e^{2c} \leq e^{2(0.1)} = e^{0.2}$ for $0 \leq c \leq 0.1$. Thus an upper bound for the local truncation error is $0.001333e^{0.2} = 0.001628$.

(c) Since $y(0.1) = e^{0.2} = 1.221403$, the actual error is $y(0.1) - y_1 = 0.001403$ which is less than 0.001628.

(d) Using the improved Euler method with $h = 0.05$ we obtain $y(0.1) \approx y_2 = 1.221025$.

(e) The error in (d) is $1.221403 - 1.221025 = 0.000378$. With global truncation error $O(h^2)$, when the step size is halved we expect the error for $h = 0.05$ to be one-fourth the error for $h = 0.1$. Comparing 0.000378 with 0.001403 we see that this is the case.

21. **(a)** Using $y'' = 38e^{-3(x-1)}$ we see that the local truncation error is

$$y''(c)\,\frac{h^2}{2} = 38e^{-3(c-1)}\,\frac{h^2}{2} = 19h^2e^{-3(c-1)}.$$

(b) Since $e^{-3(x-1)}$ is a decreasing function for $1 \le x \le 1.5$, $e^{-3(c-1)} \le e^{-3(1-1)} = 1$ for $1 \le c \le 1.5$ and

$$y''(c)\,\frac{h^2}{2} \le 19(0.1)^2(1) = 0.19.$$

(c) Using the Euler method with $h = 0.1$ we obtain $y(1.5) \approx 1.8207$. With $h = 0.05$ we obtain $y(1.5) \approx 1.9424$.

(d) Since $y(1.5) = 2.0532$, the error for $h = 0.1$ is $E_{0.1} = 0.2325$, while the error for $h = 0.05$ is $E_{0.05} = 0.1109$. With global truncation error $O(h)$ we expect $E_{0.1}/E_{0.05} \approx 2$. We actually have $E_{0.1}/E_{0.05} = 2.10$.

24. **(a)** Using $y''' = \dfrac{2}{(x+1)^3}$ we see that the local truncation error is

$$y'''(c)\,\frac{h^3}{6} = \frac{1}{(c+1)^3}\,\frac{h^3}{3}.$$

(b) Since $\dfrac{1}{(x+1)^3}$ is a decreasing function for $0 \le x \le 0.5$, $\dfrac{1}{(c+1)^3} \le \dfrac{1}{(0+1)^3} = 1$ for $0 \le c \le 0.5$ and

$$y'''(c)\,\frac{h^3}{6} \le (1)\,\frac{(0.1)^3}{3} = 0.000333.$$

(c) Using the improved Euler method with $h = 0.1$ we obtain $y(0.5) \approx 0.405281$. With $h = 0.05$ we obtain $y(0.5) \approx 0.405419$.

(d) Since $y(0.5) = 0.405465$, the error for $h = 0.1$ is $E_{0.1} = 0.000184$, while the error for $h = 0.05$ is $E_{0.05} = 0.000046$. With global truncation error $O(h^2)$ we expect $E_{0.1}/E_{0.05} \approx 4$. We actually have $E_{0.1}/E_{0.05} = 3.98$.

Exercises 9.3

3.

x(n)	y(n)
1.00	5.0000
1.10	3.9724
1.20	3.2284
1.30	2.6945
1.40	2.3163
1.50	2.0533

6.

x(n)	y(n)
0.00	1.0000
0.10	1.1115
0.20	1.2530
0.30	1.4397
0.40	1.6961
0.50	2.0670

9.

x(n)	y(n)
0.00	0.5000
0.10	0.5213
0.20	0.5358
0.30	0.5443
0.40	0.5482
0.50	0.5493

12.

x(n)	y(n)
0.00	0.5000
0.10	0.5250
0.20	0.5498
0.30	0.5744
0.40	0.5987
0.50	0.6225

15. (a)

	h = 0.05	h = 0.1
x(n)	*y(n)*	*y(n)*
1.00	1.0000	1.0000
1.05	1.1112	
1.10	1.2511	1.2511
1.15	1.4348	
1.20	1.6934	1.6934
1.25	2.1047	
1.30	2.9560	2.9425
1.35	7.8981	
1.40	1.06E+15	903.0282

(b)

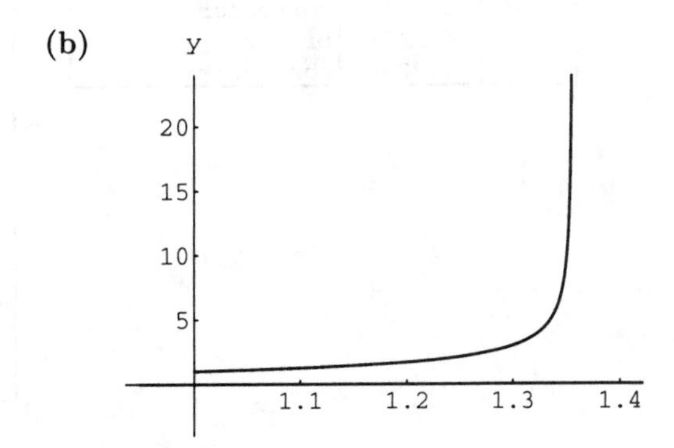

18. (a) Using $y^{(5)} = -1026e^{-3(x-1)}$ we see that the local truncation error is

$$\left| y^{(5)}(c)\,\frac{h^5}{120} \right| = 8.55h^5 e^{-3(c-1)}.$$

(b) Since $e^{-3(x-1)}$ is a decreasing function for $1 \le x \le 1.5$, $e^{-3(c-1)} \le e^{-3(1-1)} = 1$ for $1 \le c \le 1.5$ and

$$y^{(5)}(c)\,\frac{h^5}{120} \le 8.55(0.1)^5(1) = 0.0000855.$$

(c) Using the fourth-order Runge-Kutta method with $h = 0.1$ we obtain $y(1.5) \approx 2.053338827$. With $h = 0.05$ we obtain $y(1.5) \approx 2.053222989$.

Exercises 9.4

3.

x(n)	y(n)	
0.00	1.0000	initial condition
0.20	0.7328	Runge-Kutta
0.40	0.6461	Runge-Kutta
0.60	0.6585	Runge-Kutta
	0.7332	*predictor*
0.80	0.7232	corrector

6.

x(n)	y(n)	
0.00	1.0000	initial condition
0.20	1.4414	Runge-Kutta
0.40	1.9719	Runge-Kutta
0.60	2.6028	Runge-Kutta
	3.3483	*predictor*
0.80	3.3486	corrector
	4.2276	*predictor*
1.00	4.2280	corrector

x(n)	y(n)	
0.00	1.0000	initial condition
0.10	1.2102	Runge-Kutta
0.20	1.4414	Runge-Kutta
0.30	1.6949	Runge-Kutta
	1.9719	*predictor*
0.40	1.9719	corrector
	2.2740	*predictor*
0.50	2.2740	corrector
	2.6028	*predictor*
0.60	2.6028	corrector
	2.9603	*predictor*
0.70	2.9603	corrector
	3.3486	*predictor*
0.80	3.3486	corrector
	3.7703	*predictor*
0.90	3.7703	corrector
	4.2280	*predictor*
1.00	4.2280	corrector

Exercises 9.5

3. The substitution $y' = u$ leads to the system

$$y' = u, \qquad u' = 4u - 4y.$$

Using formulas (5) and (6) in the text with x corresponding to t, y corresponding to x, and u corresponding to y, we obtain

Runge-Kutta method with h=0.2

m1	m2	m3	m4	k1	k2	k3	k4	x	y	u
								0.00	-2.0000	1.0000
0.2000	0.4400	0.5280	0.9072	2.4000	3.2800	3.5360	4.8064	0.20	-1.4928	4.4731

Runge-Kutta method with h=0.1

m1	m2	m3	m4	k1	k2	k3	k4	x	y	u
								0.00	-2.0000	1.0000
0.1000	0.1600	0.1710	0.2452	1.2000	1.4200	1.4520	1.7124	0.10	-1.8321	2.4427
0.2443	0.3298	0.3444	0.4487	1.7099	2.0031	2.0446	2.3900	0.20	-1.4919	4.4753

6.

Runge-Kutta method with h=0.1

m1	m2	m3	m4	k1	k2	k3	k4	t	i1	i2
								0.00	0.0000	0.0000
10.0000	0.0000	12.5000	-20.0000	0.0000	5.0000	-5.0000	22.5000	0.10	2.5000	3.7500
8.7500	-2.5000	13.4375	-28.7500	-5.0000	4.3750	-10.6250	29.6875	0.20	2.8125	5.7813
10.1563	-4.3750	17.0703	-40.0000	-8.7500	5.0781	-16.0156	40.3516	0.30	2.0703	7.4023
13.2617	-6.3672	22.9443	-55.1758	-12.7344	6.6309	-22.5488	55.3076	0.40	0.6104	9.1919
17.9712	-8.8867	31.3507	-75.9326	-17.7734	8.9856	-31.2024	75.9821	0.50	-1.5619	11.4877

9.

Runge-Kutta method with h=0.2

m1	m2	m3	m4	k1	k2	k3	k4	t	x	y
								0.00	-3.0000	5.0000
-1.0000	-0.9200	-0.9080	-0.8176	-0.6000	-0.7200	-0.7120	-0.8216	0.20	-3.9123	4.2857

Runge-Kutta method with h=0.1

m1	m2	m3	m4	k1	k2	k3	k4	t	x	y
								0.00	-3.0000	5.0000
-0.5000	-0.4800	-0.4785	-0.4571	-0.3000	-0.3300	-0.3290	-0.3579	0.10	-3.4790	4.6707
-0.4571	-0.4342	-0.4328	-0.4086	-0.3579	-0.3858	-0.3846	-0.4112	0.20	-3.9123	4.2857

12. Solving for x' and y' we obtain the system

$$x' = \frac{1}{2}y - 3t^2 + 2t - 5$$

$$y' = -\frac{1}{2}y + 3t^2 + 2t + 5.$$

Exercises 9.5

Runge-Kutta method with h=0.2

m1	m2	m3	m4	k1	k2	k3	k4	t	x	y
								0.00	3.0000	-1.0000
-1.1000	-1.0110	-1.0115	-0.9349	1.1000	1.0910	1.0915	1.0949	0.20	1.9867	0.0933

Runge-Kutta method with h=0.1

m1	m2	m3	m4	k1	k2	k3	k4	t	x	y
								0.00	3.0000	-1.0000
-0.5500	-0.5270	-0.5271	-0.5056	0.5500	0.5470	0.5471	0.5456	0.10	2.4727	-0.4527
-0.5056	-0.4857	-0.4857	-0.4673	0.5456	0.5457	0.5457	0.5473	0.20	1.9867	0.0933

Exercises 9.6

3. We identify $P(x) = 2$, $Q(x) = 1$, $f(x) = 5x$, and $h = (1-0)/5 = 0.2$. Then the finite difference equation is

$$1.2y_{i+1} - 1.96y_i + 0.8y_{i-1} = 0.04(5x_i).$$

The solution of the corresponding linear system gives

x	0.0	0.2	0.4	0.6	0.8	1.0
y	0.0000	-0.2259	-0.3356	-0.3308	-0.2167	0.0000

6. We identify $P(x) = 5$, $Q(x) = 0$, $f(x) = 4\sqrt{x}$, and $h = (2-1)/6 = 0.1667$. Then the finite difference equation is

$$1.4167y_{i+1} - 2y_i + 0.5833y_{i-1} = 0.2778(4\sqrt{x_i}).$$

The solution of the corresponding linear system gives

x	1.0000	1.1667	1.3333	1.5000	1.6667	1.8333	2.0000
y	1.0000	-0.5918	-1.1626	-1.3070	-1.2704	-1.1541	-1.0000

9. We identify $P(x) = 1 - x$, $Q(x) = x$, $f(x) = x$, and $h = (1-0)/10 = 0.1$. Then the finite difference equation is

$$[1 + 0.05(1 - x_i)]y_{i+1} + [-2 + 0.01x_i]y_i + [1 - 0.05(1 - x_i)]y_{i-1} = 0.01x_i.$$

The solution of the corresponding linear system gives

x	0.0	0.1	0.2	0.3	0.4	0.5	0.6
y	0.0000	0.2660	0.5097	0.7357	0.9471	1.1465	1.3353

0.7	0.8	0.9	1.0
1.5149	1.6855	1.8474	2.0000

12. We identify $P(r) = 2/r$, $Q(r) = 0$, $f(r) = 0$, and $h = (4-1)/6 = 0.5$. Then the finite difference equation is

$$\left(1 + \frac{0.5}{r_i}\right)u_{i+1} - 2u_i + \left(1 - \frac{0.5}{r_i}\right)u_{i-1} = 0.$$

The solution of the corresponding linear system gives

r	1.0	1.5	2.0	2.5	3.0	3.5	4.0
u	50.0000	72.2222	83.3333	90.0000	94.4444	97.6190	100.0000

Chapter 9 Review Exercises

3.

h=0.1		IMPROVED	RUNGE
x(n)	EULER	EULER	KUTTA
1.00	2.0000	2.0000	2.0000
1.10	2.1386	2.1549	2.1556
1.20	2.3097	2.3439	2.3454
1.30	2.5136	2.5672	2.5695
1.40	2.7504	2.8246	2.8278
1.50	3.0201	3.1157	3.1197

h=0.05		IMPROVED	RUNGE
x(n)	EULER	EULER	KUTTA
1.00	2.0000	2.0000	2.0000
1.05	2.0693	2.0735	2.0736
1.10	2.1469	2.1554	2.1556
1.15	2.2328	2.2459	2.2462
1.20	2.3272	2.3450	2.3454
1.25	2.4299	2.4527	2.4532
1.30	2.5409	2.5689	2.5695
1.35	2.6604	2.6937	2.6944
1.40	2.7883	2.8269	2.8278
1.45	2.9245	2.9686	2.9696
1.50	3.0690	3.1187	3.1197

6.

h=0.1		IMPROVED	RUNGE
x(n)	EULER	EULER	KUTTA
1.00	1.0000	1.0000	1.0000
1.10	1.2000	1.2380	1.2415
1.20	1.4760	1.5910	1.6036
1.30	1.8710	2.1524	2.1909
1.40	2.4643	3.1458	3.2745
1.50	3.4165	5.2510	5.8338

h=0.05		IMPROVED	RUNGE
x(n)	EULER	EULER	KUTTA
1.00	1.0000	1.0000	1.0000
1.05	1.1000	1.1091	1.1095
1.10	1.2183	1.2405	1.2415
1.15	1.3595	1.4010	1.4029
1.20	1.5300	1.6001	1.6036
1.25	1.7389	1.8523	1.8586
1.30	1.9988	2.1799	2.1911
1.35	2.3284	2.6197	2.6401
1.40	2.7567	3.2360	3.2755
1.45	3.3296	4.1528	4.2363
1.50	4.1253	5.6404	5.8446

9. Using $x_0 = 1$, $y_0 = 2$, and $h = 0.1$ we have

$$x_1 = x_0 + h(x_0 + y_0) = 1 + 0.1(1 + 2) = 1.3$$

$$y_1 = y_0 + h(x_0 - y_0) = 2 + 0.1(1 - 2) = 1.9$$

and

$$x_2 = x_1 + h(x_1 + y_1) = 1.3 + 0.1(1.3 + 1.9) = 1.62$$

$$y_2 = y_1 + h(x_1 - y_1) = 1.9 + 0.1(1.3 - 1.9) = 1.84.$$

Thus, $x(0.2) \approx 1.62$ and $y(0.2) \approx 1.84$.

10 Plane Autonomous Systems and Stability

Exercises 10.1

3. The corresponding plane autonomous system is

$$x' = y, \quad y' = x^2 - y(1 - x^3).$$

If (x, y) is a critical point, $y = 0$ and so $x^2 - y(1 - x^3) = x^2 = 0$. Therefore $(0, 0)$ is the sole critical point.

6. The corresponding plane autonomous system is

$$x' = y, \quad y' = -x + \epsilon x|x|.$$

If (x, y) is a critical point, $y = 0$ and $-x + \epsilon x|x| = x(-1 + \epsilon|x|) = 0$. Hence $x = 0,\ 1/\epsilon,\ -1/\epsilon$. The critical points are $(0, 0)$, $(1/\epsilon, 0)$ and $(-1/\epsilon, 0)$.

9. From $x - y = 0$ we have $y = x$. Substituting into $3x^2 - 4y = 0$ we obtain $3x^2 - 4x = x(3x - 4) = 0$. It follows that $(0, 0)$ and $(4/3, 4/3)$ are the critical points of the system.

12. Adding the two equations we obtain $10 - 15\dfrac{y}{y+5} = 0$. It follows that $y = 10$, and from $-2x + y + 10 = 0$ we may conclude that $x = 10$. Therefore $(10, 10)$ is the sole critical point of the system.

15. From $x(1 - x^2 - 3y^2) = 0$ we have $x = 0$ or $x^2 + 3y^2 = 1$. If $x = 0$, then substituting into $y(3 - x^2 - 3y^2)$ gives $y(3 - 3y^2) = 0$. Therefore $y = 0,\ 1,\ -1$. Likewise $x^2 = 1 - 3y^2$ yields $2y = 0$ so that $y = 0$ and $x^2 = 1 - 3(0)^2 = 1$. The critical points of the system are therefore $(0, 0)$, $(0, 1)$, $(0, -1)$, $(1, 0)$, and $(-1, 0)$.

18. (a) From Exercises 8.2, Problem 6, $x = c_1 + 2c_2e^{-5t}$ and $y = 3c_1 + c_2e^{-5t}$.

(b) From $\mathbf{X}(0) = (3, 4)$ it follows that $c_1 = c_2 = 1$. Therefore $x = 1 + 2e^{-5t}$ and $y = 3 + e^{-5t}$.

(c)

21. **(a)** From Exercises 8.2, Problem 33, $x = c_1(\sin t - \cos t)e^{4t} + c_2(-\sin t - \cos t)e^{4t}$ and $y = 2c_1(\cos t)\,e^{4t} + 2c_2(\sin t)\,e^{4t}$. Because of the presence of e^{4t}, there are no periodic solutions.

(b) From $\mathbf{X}(0) = (-1, 2)$ it follows that $c_1 = 1$ and $c_2 = 0$. Therefore $x = (\sin t - \cos t)e^{4t}$ and $y = 2(\cos t)\,e^{4t}$.

(c)

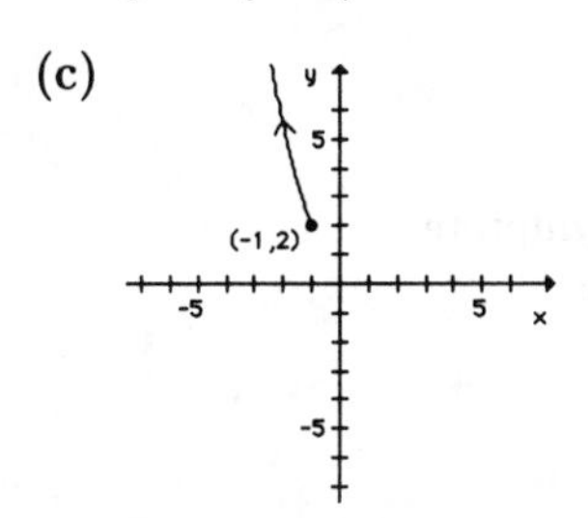

24. Switching to polar coordinates,

$$\frac{dr}{dt} = \frac{1}{r}\left(x\,\frac{dx}{dt} + y\,\frac{dy}{dt}\right) = \frac{1}{r}(xy - x^2r^2 - xy + y^2r^2) = r^3$$

$$\frac{d\theta}{dt} = \frac{1}{r^2}\left(-y\,\frac{dx}{dt} + x\,\frac{dy}{dt}\right) = \frac{1}{r^2}(-y^2 - xyr^2 - x^2 + xyr^2) = -1\,.$$

If we use separation of variables, it follows that

$$r = \frac{1}{\sqrt{-2t + c_1}} \quad \text{and} \quad \theta = -t + c_2.$$

Since $\mathbf{X}(0) = (4, 0)$, $r = 4$ and $\theta = 0$ when $t = 0$. It follows that $c_2 = 0$ and $c_1 = \frac{1}{16}$. The final solution may be written as

$$r = \frac{4}{\sqrt{1 - 32t}}, \qquad \theta = -t.$$

Note that $r \to \infty$ as $t \to \left(\frac{1}{32}\right)^-$. Because $0 \le t \le \frac{1}{32}$, the curve is not a spiral.

27. The system has no critical points, so there are no periodic solutions.

30. The system has no critical points, so there are no periodic solutions.

Exercises 10.2

3. **(a)** All solutions are unstable spirals which become unbounded as t increases.

(b)

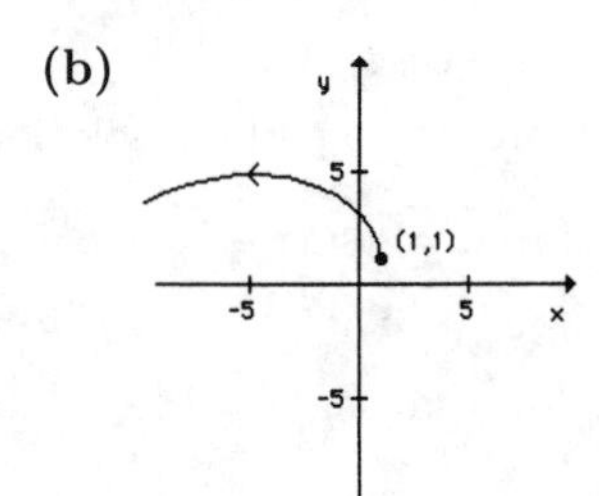

6. (a) All solutions become unbounded and $y = x/2$ serves as the asymptote.

(b)

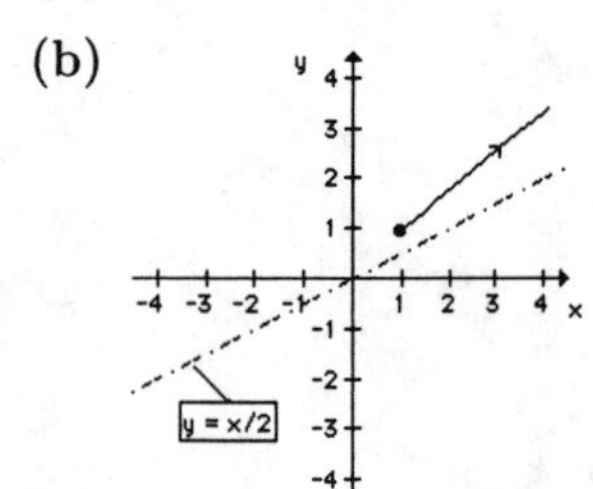

9. Since $\Delta = -41 < 0$, we may conclude from Figure 10.18 that $(0,0)$ is a saddle point.

12. Since $\Delta = 1$ and $\tau = -1$, $\tau^2 - 4\Delta = -3$ and so from Figure 10.18, $(0,0)$ is a stable spiral point.

15. Since $\Delta = 0.01$ and $\tau = -0.03$, $\tau^2 - 4\Delta < 0$ and so from Figure 10.18, $(0,0)$ is a stable spiral point.

18. Note that $\Delta = 1$ and $\tau = \mu$. Therefore we need both $\tau = \mu < 0$ and $\tau^2 - 4\Delta = \mu^2 - 4 < 0$ for $(0,0)$ to be a stable spiral point. These two conditions may be written as $-2 < \mu < 0$.

21. $\mathbf{AX}_1 + \mathbf{F} = \mathbf{0}$ implies that $\mathbf{AX}_1 = -\mathbf{F}$ or $\mathbf{X}_1 = -\mathbf{A}^{-1}\mathbf{F}$. Since $\mathbf{X}_p(t) = -\mathbf{A}^{-1}\mathbf{F}$ is a particular solution, it follows from Theorem 8.6 that $\mathbf{X}(t) = \mathbf{X}_c(t) + \mathbf{X}_1$ is the general solution to $\mathbf{X}' = \mathbf{AX} + \mathbf{F}$. If $\tau < 0$ and $\Delta > 0$ then $\mathbf{X}_c(t)$ approaches $(0,0)$ by Theorem 10.1(a). It follows that $\mathbf{X}(t)$ approaches $\mathbf{X}_1$ as $t \to \infty$.

24. (a) The critical point is $\mathbf{X}_1 = (-1, -2)$.

(b) From the graph, $\mathbf{X}_1$ appears to be a stable node or a degenerate stable node.

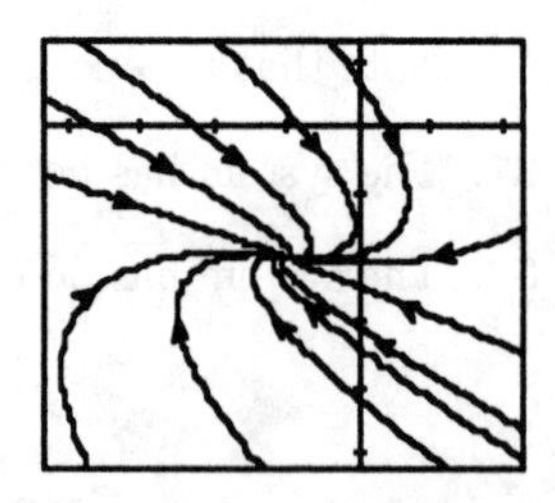

(c) Since $\tau = -16$, $\Delta = 64$, and $\tau^2 - 4\Delta = 0$, $(0,0)$ is a degenerate stable node.

Exercises 10.3

3. The critical points are $x = 0$ and $x = n+1$. Since $g'(x) = k(n+1) - 2kx$, $g'(0) = k(n+1) > 0$ and $g'(n+1) = -k(n+1) < 0$. Therefore $x = 0$ is unstable while $x = n+1$ is asymptotically stable. See Theorem 10.2.

6. The only critical point is $v = mg/k$. Now $g(v) = g - (k/m)v$ and so $g'(v) = -k/m < 0$. Therefore $v = mg/k$ is an asymptotically stable critical point by Theorem 10.2.

9. Critical points occur at $P = a/b$, c but not at $P = 0$. Since $g'(P) = (a - bP) + (P - c)(-b)$,

$$g'(a/b) = (a/b - c)(-b) = -a + bc \quad \text{and} \quad g'(c) = a - bc.$$

Since $a < bc$, $-a + bc > 0$ and $a - bc < 0$. Therefore $P = a/b$ is unstable while $P = c$ is asymptotically stable.

12. Critical points are $(1,0)$ and $(-1,0)$, and

$$\mathbf{g}'(\mathbf{X}) = \begin{pmatrix} 2x & -2y \\ 0 & 2 \end{pmatrix}.$$

At $\mathbf{X} = (1,0)$, $\tau = 4$, $\Delta = 4$, and so $\tau^2 - 4\Delta = 0$. We may conclude that $(1,0)$ is unstable but we are unable to classify this critical point any further. At $\mathbf{X} = (-1,0)$, $\Delta = -4 < 0$ and so $(-1,0)$ is a saddle point.

15. Since $x^2 - y^2 = 0$, $y^2 = x^2$ and so $x^2 - 3x + 2 = (x-1)(x-2) = 0$. It follows that the critical points are $(1,1)$, $(1,-1)$, $(2,2)$, and $(2,-2)$. We next use the Jacobian

$$\mathbf{g}'(\mathbf{X}) = \begin{pmatrix} -3 & 2y \\ 2x & -2y \end{pmatrix}$$

to classify these four critical points. For $\mathbf{X} = (1,1)$, $\tau = -5$, $\Delta = 2$, and so $\tau^2 - 4\Delta = 17 > 0$. Therefore $(1,1)$ is a stable node. For $\mathbf{X} = (1,-1)$, $\Delta = -2 < 0$ and so $(1,-1)$ is a saddle point. For $\mathbf{X} = (2,2)$, $\Delta = -4 < 0$ and so we have another saddle point. Finally, if $\mathbf{X} = (2,-2)$, $\tau = 1$, $\Delta = 4$, and so $\tau^2 - 4\Delta = -15 < 0$. Therefore $(2,-2)$ is an unstable spiral point.

18. We found that $(0,0)$, $(0,1)$, $(0,-1)$, $(1,0)$ and $(-1,0)$ were the critical points in Exercise 15, Section 10.1. The Jacobian is

$$\mathbf{g}'(\mathbf{X}) = \begin{pmatrix} 1 - 3x^2 - 3y^2 & -6xy \\ -2xy & 3 - x^2 - 9y^2 \end{pmatrix}.$$

For $\mathbf{X} = (0,0)$, $\tau = 4$, $\Delta = 3$ and so $\tau^2 - 4\Delta = 4 > 0$. Therefore $(0,0)$ is an unstable node. Both $(0,1)$ and $(0,-1)$ give $\tau = -8$, $\Delta = 12$, and $\tau^2 - 4\Delta = 16 > 0$. These two critical points are therefore stable nodes. For $\mathbf{X} = (1,0)$ or $(-1,0)$, $\Delta = -4 < 0$ and so saddle points occur.

21. The corresponding plane autonomous system is

$$\theta' = y, \quad y' = (\cos\theta - \frac{1}{2})\sin\theta.$$

Since $|\theta| < \pi$, it follows that critical points are $(0,0)$, $(\pi/3,0)$ and $(-\pi/3,0)$. The Jacobian matrix is

$$\mathbf{g}'(\mathbf{X}) = \begin{pmatrix} 0 & 1 \\ \cos 2\theta - \frac{1}{2}\cos\theta & 0 \end{pmatrix}$$

and so at $(0,0)$, $\tau = 0$ and $\Delta = -1/2$. Therefore $(0,0)$ is a saddle point. For $\mathbf{X} = (\pm\pi/3, 0)$, $\tau = 0$ and $\Delta = 3/4$. It is not possible to classify either critical point in this borderline case.

24. The corresponding plane autonomous system is

$$x' = y, \quad y' = -\frac{4x}{1+x^2} - 2y$$

and the only critical point is $(0,0)$. Since the Jacobian matrix is

$$\mathbf{g}'(\mathbf{X}) = \begin{pmatrix} 0 & 1 \\ -4\frac{1-x^2}{(1+x^2)^2} & -2 \end{pmatrix},$$

$\tau = -2$, $\Delta = 4$, $\tau^2 - 4\Delta = -12$, and so $(0,0)$ is a stable spiral point.

27. The corresponding plane autonomous system is

$$x' = y, \quad y' = -\frac{(\beta + \alpha^2 y^2)x}{1 + \alpha^2 x^2}$$

and the Jacobian matrix is

$$\mathbf{g}'(\mathbf{X}) = \begin{pmatrix} 0 & 1 \\ \frac{(\beta+\alpha y^2)(\alpha^2 x^2 - 1)}{(1+\alpha^2 x^2)^2} & \frac{-2\alpha^2 yx}{1+\alpha^2 x^2} \end{pmatrix}.$$

For $\mathbf{X} = (0,0)$, $\tau = 0$ and $\Delta = \beta$. Since $\beta < 0$, we may conclude that $(0,0)$ is a saddle point.

30. **(a)** The corresponding plane autonomous system is

$$x' = y, \quad y' = \epsilon(y - \frac{1}{3}y^3) - x$$

and so the only critical point is $(0,0)$. Since the Jacobian matrix is

$$\mathbf{g}'(\mathbf{X}) = \begin{pmatrix} 0 & 1 \\ -1 & \epsilon(1 - y^2) \end{pmatrix},$$

$\tau = \epsilon$, $\Delta = 1$, and so $\tau^2 - 4\Delta = \epsilon^2 - 4$ at the critical point $(0,0)$.

(b) When $\tau = \epsilon > 0$, $(0,0)$ is an unstable critical point.

(c) When $\epsilon < 0$ and $\tau^2 - 4\Delta = \epsilon^2 - 4 < 0$, $(0,0)$ is a stable spiral point. These two requirements can be written as $-2 < \epsilon < 0$.

(d) When $\epsilon = 0$, $x'' + x = 0$ and so $x = c_1 \cos t + c_2 \sin t$. Therefore all solutions are periodic (with period 2π) and so $(0, 0)$ is a center.

33. (a) $x' = 2xy = 0$ implies that either $x = 0$ or $y = 0$. If $x = 0$, then from $1 - x^2 + y^2 = 0$, $y^2 = -1$ and there are no real solutions. If $y = 0$, $1 - x^2 = 0$ and so $(1, 0)$ and $(-1, 0)$ are critical points. The Jacobian matrix is

$$\mathbf{g}'(\mathbf{X}) = \begin{pmatrix} 2y & 2x \\ -2x & 2y \end{pmatrix}$$

and so $\tau = 0$ and $\Delta = 4$ at either $\mathbf{X} = (1, 0)$ or $(-1, 0)$. We obtain no information about these critical points in this borderline case.

(b) $\dfrac{dy}{dx} = \dfrac{y'}{x'} = \dfrac{1 - x^2 + y^2}{2xy}$ or $2xy\dfrac{dy}{dx} = 1 - x^2 + y^2$. Letting $\mu = \dfrac{y^2}{x}$, it follows that $\dfrac{d\mu}{dx} = \dfrac{1}{x^2} - 1$ and so $\mu = -\dfrac{1}{x} - x + 2c$. Therefore $\dfrac{y^2}{x} = -\dfrac{1}{x} - x + 2c$ which can be put in the form

$$(x - c)^2 + y^2 = c^2 - 1.$$

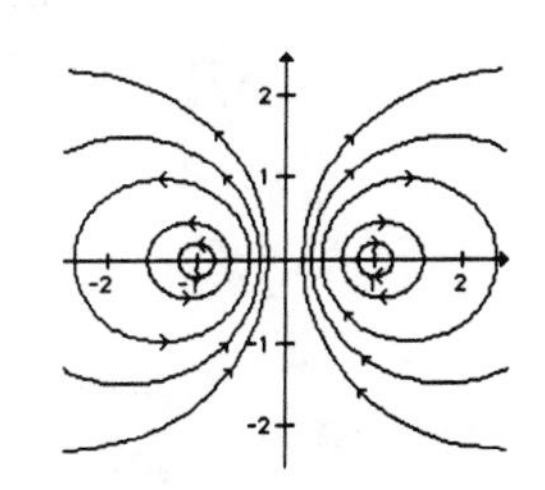

The solution curves are shown and so both $(1, 0)$ and $(-1, 0)$ are centers.

36. The corresponding plane autonomous system is

$$x' = y, \quad y' = \epsilon x^2 - x + 1$$

and so the critical points must satisfy $y = 0$ and

$$x = \frac{1 \pm \sqrt{1 - 4\epsilon}}{2\epsilon}.$$

Therefore we must require that $\epsilon \le \frac{1}{4}$ for real solutions to exist. We will use the Jacobian matrix

$$\mathbf{g}'(\mathbf{X}) = \begin{pmatrix} 0 & 1 \\ 2\epsilon x - 1 & 0 \end{pmatrix}$$

to attempt to classify $((1 \pm \sqrt{1 - 4\epsilon})/2\epsilon, 0)$ when $\epsilon \le 1/4$. Note that $\tau = 0$ and $\Delta = \mp\sqrt{1 - 4\epsilon}$. For $\mathbf{X} = ((1 + \sqrt{1 - 4\epsilon})/2\epsilon, 0)$ and $\epsilon < 1/4$, $\Delta < 0$ and so a saddle point occurs. For $\mathbf{X} = ((1 - \sqrt{1 - 4\epsilon})/2\epsilon, 0)$ $\Delta \ge 0$ and we are not able to classify this critical point using linearization.

39. (a) Letting $x = \theta$ and $y = x'$ we obtain the system $x' = y$ and $y' = \frac{1}{2} - \sin x$. Since $\sin \pi/6 = \sin 5\pi/6 = \frac{1}{2}$ we see that $(\pi/6, 0)$ and $(5\pi/6, 0)$ are critical points of the system.

(b) The Jacobian matrix is

$$\mathbf{g}'(\mathbf{X}) = \begin{pmatrix} 0 & 1 \\ -\cos x & 0 \end{pmatrix}$$

and so

$$\mathbf{A}_1 = \mathbf{g}' = ((\pi/6,0)) = \begin{pmatrix} 0 & 1 \\ -\sqrt{3}/2 & 0 \end{pmatrix} \quad \text{and} \quad \mathbf{A}_2 = \mathbf{g}' = ((5\pi/6,0)) = \begin{pmatrix} 0 & 1 \\ \sqrt{3}/2 & 0 \end{pmatrix}.$$

Since $\det \mathbf{A}_1 > 0$ and the trace of $\mathbf{A}_1$ is 0, no conclusion can be drawn regarding the critical point $(\pi/6, 0)$. Since $\det \mathbf{A}_2 < 0$, we see that $(5\pi/6, 0)$ is a saddle point.

(c) From the system in part (a) we obtain the first-order differential equation

$$\frac{dy}{dx} = \frac{1/2 - \sin x}{y}.$$

Separating variables and integrating we obtain

$$\int y\,dy = \int \left(\frac{1}{2} - \sin x\right) dx$$

and

$$\frac{1}{2}y^2 = \frac{1}{2}x + \cos x + c_1$$

or

$$y^2 = x + 2\cos x + c_2.$$

For x_0 near $\pi/6$, if $\mathbf{X}(0) = (x_0, 0)$ then $c_2 = -x_0 - 2\cos x_0$ and $y^2 = x + 2\cos x - x_0 - 2\cos x_0$. Thus, there are two values of y for each x in a sufficiently small interval around $\pi/6$. Therefore $(\pi/6, 0)$ is a center.

Exercises 10.4

3. The corresponding plane autonomous system is

$$x' = y, \quad y' = -g\frac{f'(x)}{1 + [f'(x)]^2} - \frac{\beta}{m}y$$

and

$$\frac{\partial}{\partial x}\left(-g\frac{f'(x)}{1 + [f'(x)]^2} - \frac{\beta}{m}y\right) = -g\frac{(1 + [f'(x)]^2)f''(x) - f'(x)2f'(x)f''(x)}{(1 + [f'(x)]^2)^2}.$$

If $\mathbf{X}_1 = (x_1, y_1)$ is a critical point, $y_1 = 0$ and $f'(x_1) = 0$. The Jacobian at this critical point is therefore

$$\mathbf{g}'(\mathbf{X}_1) = \begin{pmatrix} 0 & 1 \\ -gf''(x_1) & -\frac{\beta}{m} \end{pmatrix}.$$

6. **(a)** If $f(x) = \cosh x$, $f'(x) = \sinh x$ and $[f'(x)]^2 + 1 = \sinh^2 x + 1 = \cosh^2 x$. Therefore

$$\frac{dy}{dx} = \frac{y'}{x'} = -g\frac{\sinh x}{\cosh^2 x}\frac{1}{y}.$$

We may separate variables to show that $y^2 = \dfrac{2g}{\cosh x} + c$. But $x(0) = x_0$ and $y(0) = x'(0) = v_0$. Therefore $c = v_0^2 - \dfrac{2g}{\cosh x_0}$ and so

$$y^2 = \frac{2g}{\cosh x} - \frac{2g}{\cosh x_0} + v_0^2.$$

Now

$$\frac{2g}{\cosh x} - \frac{2g}{\cosh x_0} + v_0^2 \geq 0 \quad \text{if and only if} \quad \cosh x \leq \frac{2g \cosh x_0}{2g - v_0^2 \cosh x_0}$$

and the solution to this inequality is an interval $[-a, a]$. Therefore each x in $(-a, a)$ has two corresponding values of y and so the solution is periodic.

(b) Since $z = \cosh x$, the maximum height occurs at the largest value of x on the cycle. From (a), $x_{max} = a$ where $\cosh a = \dfrac{2g \cosh x_0}{2g - v_0^2 \cosh x_0}$. Therefore

$$z_{max} = \frac{2g \cosh x_0}{2g - v_0^2 \cosh x_0}.$$

9. (a) In the Lotka-Volterra Model the average number of predators is d/c and the average number of prey is a/b. But

$$x' = -ax + bxy - \epsilon_1 x = -(a + \epsilon_1)x + bxy$$

$$y' = -cxy + dy - \epsilon_2 y = -cxy + (d - \epsilon_2)y$$

and so the new critical point in the first quadrant is $(d/c - \epsilon_2/c, a/b + \epsilon_1/b)$.

(b) The average number of predators $d/c - \epsilon_2/c$ has decreased while the average number of prey $a/b + \epsilon_1/b$ has increased. The fishery science model is consistent with Volterra's principle.

12. $\Delta = r_1 r_2$, $\tau = r_1 + r_2$ and $\tau^2 - 4\Delta = (r_1 + r_2)^2 - 4r_1 r_2 = (r_1 - r_2)^2$. Therefore when $r_1 \neq r_2$, $(0, 0)$ is an unstable node.

15. $\dfrac{K_1}{\alpha_{12}} < K_2 < K_1 \alpha_{21}$ and so $\alpha_{12}\alpha_{21} > 1$. Therefore $\Delta = (1 - \alpha_{12}\alpha_{21})\hat{x}\hat{y}\dfrac{r_1 r_2}{K_1 K_2} < 0$ and so $(\hat{x}, \hat{y})$ is a saddle point.

Chapter 10 Review Exercises

3. a center or a saddle point

6. True

9. The system is linear and we identify $\Delta = -\alpha$ and $\tau = \alpha + 1$. Since a critical point will be a center when $\Delta > 0$ and $\tau = 0$ we see that for $\alpha = -1$ critical points will be centers and solutions will be periodic. Note also that when $\alpha = -1$ the system is

$$x' = -x - 2y$$

$$y' = x + y,$$

which does have an isolated critical point at $(0, 0)$.

12. (a) If $\mathbf{X}(0) = \mathbf{X}_0$ lies on the line $y = -2x$, then $\mathbf{X}(t)$ approaches $(0, 0)$ along this line. For all other initial conditions, $\mathbf{X}(t)$ approaches $(0, 0)$ from the direction determined by the line $y = x$.

(b) If $\mathbf{X}(0) = \mathbf{X}_0$ lies on the line $y = -x$, then $\mathbf{X}(t)$ approaches $(0, 0)$ along this line. For all other initial conditions, $\mathbf{X}(t)$ becomes unbounded and $y = 2x$ serves as an asymptote.

15. From $x = r\cos\theta$, $y = r\sin\theta$ we have

$$\frac{dx}{dt} = -r\sin\theta\,\frac{d\theta}{dt} + \frac{dr}{dt}\cos\theta$$

$$\frac{dy}{dt} = r\cos\theta\,\frac{d\theta}{dt} + \frac{dr}{dt}\sin\theta.$$

Then $r' = \alpha r$, $\theta' = 1$ gives

$$\frac{dx}{dt} = -r\sin\theta + \alpha r\cos\theta$$

$$\frac{dy}{dt} = r\cos\theta + \alpha r\sin\theta.$$

We see that $r = 0$, which corresponds to $\mathbf{X} = (0, 0)$, is a critical point. Solving $r' = \alpha r$ we have $r = c_1 e^{\alpha t}$. Thus, when $\alpha < 0$, $\lim_{t\to\infty} r(t) = 0$ and $(0, 0)$ is a stable critical point. When $\alpha = 0$, $r' = 0$ and $r = c_1$. In this case $(0, 0)$ is a center, which is stable. Therefore, $(0, 0)$ is a stable critical point for the system when $\alpha \le 0$.

18. $\dfrac{dy}{dx} = \dfrac{y'}{x'} = \dfrac{-2x\sqrt{y^2+1}}{y}$. We may separate variables to show that $\sqrt{y^2+1} = -x^2 + c$. But $x(0) = x_0$ and $y(0) = x'(0) = 0$. It follows that $c = 1 + x_0^2$ so that

$$y^2 = (1 + x_0^2 - x^2)^2 - 1.$$

Note that $1 + x_0^2 - x^2 > 1$ for $-x_0 < x < x_0$ and $y = 0$ for $x = \pm x_0$. Each x with $-x_0 < x < x_0$ has two corresponding values of y and so the solution $\mathbf{X}(t)$ with $\mathbf{X}(0) = (x_0, 0)$ is periodic.

11 Orthogonal Functions and Fourier Series

Exercises 11.1

3. $\displaystyle\int_0^2 e^x(xe^{-x} - e^{-x})dx = \int_0^2 (x-1)dx = \left(\frac{1}{2}x^2 - x\right)\Big|_0^2 = 0$

6. $\displaystyle\int_{\pi/4}^{5\pi/4} e^x \sin x\,dx = \left(\frac{1}{2}e^x \sin x - \frac{1}{2}e^x \cos x\right)\Big|_{\pi/4}^{5\pi/4} = 0$

9. For $m \neq n$

$$\begin{aligned}\int_0^\pi \sin nx \sin mx\,dx &= \frac{1}{2}\int_0^\pi [\cos(n-m)x - \cos(n+m)x]\,dx \\ &= \frac{1}{2(n-m)}\sin(n-m)x\Big|_0^\pi - \frac{1}{2(n+m)}\sin 2(n+m)x\Big|_0^\pi \\ &= 0.\end{aligned}$$

For $m = n$

$$\int_0^\pi \sin^2 nx\,dx = \int_0^\pi \left[\frac{1}{2} - \frac{1}{2}\cos 2nx\right]dx = \frac{1}{2}x\Big|_0^\pi - \frac{1}{4n}\sin 2nx\Big|_0^\pi = \frac{\pi}{2}$$

so that

$$\|\sin nx\| = \sqrt{\frac{\pi}{2}}\,.$$

12. For $m \neq n$, we use Problems 11 and 10:

$$\int_{-p}^p \cos\frac{n\pi}{p}x \cos\frac{m\pi}{p}x\,dx = 2\int_0^p \cos\frac{n\pi}{p}x \cos\frac{m\pi}{p}x\,dx = 0$$

$$\int_{-p}^p \sin\frac{n\pi}{p}x \sin\frac{m\pi}{p}x\,dx = 2\int_0^p \sin\frac{n\pi}{p}x \sin\frac{m\pi}{p}x\,dx = 0.$$

Also

$$\int_{-p}^p \sin\frac{n\pi}{p}x \cos\frac{m\pi}{p}x\,dx = \frac{1}{2}\int_{-p}^p \left(\sin\frac{(n-m)\pi}{p}x + \sin\frac{(n+m)\pi}{p}x\right)dx = 0,$$

$$\int_{-p}^p 1\cdot\cos\frac{n\pi}{p}x\,dx = \frac{p}{n\pi}\sin\frac{n\pi}{p}x\Big|_{-p}^p = 0,$$

$$\int_{-p}^p 1\cdot\sin\frac{n\pi}{p}x\,dx = -\frac{p}{n\pi}\cos\frac{n\pi}{p}x\Big|_{-p}^p = 0,$$

and

$$\int_{-p}^{p} \sin\frac{n\pi}{p}x \cos\frac{n\pi}{p}x\,dx = \int_{-p}^{p} \frac{1}{2}\sin\frac{2n\pi}{p}x\,dx = -\frac{p}{4n\pi}\cos\frac{2n\pi}{p}x\Big|_{-p}^{p} = 0.$$

For $m = n$

$$\int_{-p}^{p} \cos^2\frac{n\pi}{p}x\,dx = \int_{-p}^{p}\left(\frac{1}{2}+\frac{1}{2}\cos\frac{2n\pi}{p}x\right)dx = p,$$

$$\int_{-p}^{p} \sin^2\frac{n\pi}{p}x\,dx = \int_{-p}^{p}\left(\frac{1}{2}-\frac{1}{2}\cos\frac{2n\pi}{p}x\right)dx = p,$$

and

$$\int_{-p}^{p} 1^2 dx = 2p$$

so that

$$\|1\| = \sqrt{2p}\,, \quad \left\|\cos\frac{n\pi}{p}x\right\| = \sqrt{p}\,, \quad \text{and} \quad \left\|\sin\frac{n\pi}{p}x\right\| = \sqrt{p}\,.$$

15. By orthogonality $\int_a^b \phi_0(x)\phi_n(x)dx = 0$ for $n = 1, 2, 3, \ldots$; that is, $\int_a^b \phi_n(x)dx = 0$ for $n = 1, 2, 3, \ldots$.

18. Setting

$$0 = \int_{-2}^{2} f_3(x)f_1(x)\,dx = \int_{-2}^{2}\left(x^2 + c_1x^3 + c_2x^4\right)dx = \frac{16}{3} + \frac{64}{5}c_2$$

and

$$0 = \int_{-2}^{2} f_3(x)f_2(x)\,dx = \int_{-2}^{2}\left(x^3 + c_1x^4 + c_2x^5\right)dx = \frac{64}{5}c_1$$

we obtain $c_1 = 0$ and $c_2 = -5/12$.

Exercises 11.2

3. $a_0 = \int_{-1}^{1} f(x)\,dx = \int_{-1}^{0} 1\,dx + \int_0^1 x\,dx = \frac{3}{2}$

$$a_n = \int_{-1}^{1} f(x)\cos n\pi x\,dx = \int_{-1}^{0}\cos n\pi x\,dx + \int_0^1 x\cos n\pi x\,dx = \frac{1}{n^2\pi^2}[(-1)^n - 1]$$

$$b_n = \int_{-1}^{1} f(x)\sin n\pi x\,dx = \int_{-1}^{0}\sin n\pi x\,dx + \int_0^1 x\sin n\pi x\,dx = -\frac{1}{n\pi}$$

$$f(x) = \frac{3}{4} + \sum_{n=1}^{\infty}\left[\frac{(-1)^n - 1}{n^2\pi^2}\cos n\pi x - \frac{1}{n\pi}\sin n\pi x\right]$$

6. $a_0 = \frac{1}{\pi}\int_{-\pi}^{\pi} f(x)\,dx = \frac{1}{\pi}\int_{-\pi}^{0}\pi^2\,dx + \frac{1}{\pi}\int_0^{\pi}\left(\pi^2 - x^2\right)dx = \frac{5}{3}\pi^2$

$$a_n = \frac{1}{\pi}\int_{-\pi}^{\pi} f(x)\cos nx\,dx = \frac{1}{\pi}\int_{-\pi}^{0} \pi^2\cos nx\,dx + \frac{1}{\pi}\int_0^{\pi}\left(\pi^2 - x^2\right)\cos nx\,dx$$

$$= \frac{1}{\pi}\left(\frac{\pi^2 - x^2}{n}\sin nx\,\Big|_0^{\pi} + \frac{2}{n}\int_0^{\pi} x\sin nx\,dx\right) = \frac{2}{n^2}(-1)^{n+1}$$

$$b_n = \frac{1}{\pi}\int_{-\pi}^{\pi} f(x)\sin nx\,dx = \frac{1}{\pi}\int_{-\pi}^{0} \pi^2\sin nx\,dx + \frac{1}{\pi}\int_0^{\pi}\left(\pi^2 - x^2\right)\sin nx\,dx$$

$$= \frac{\pi}{n}[(-1)^n - 1] + \frac{1}{\pi}\left(\frac{x^2 - \pi^2}{n}\cos nx\,\Big|_0^{\pi} - \frac{2}{n}\int_0^{\pi} x\cos nx\,dx\right) = \frac{\pi}{n}(-1)^n + \frac{2}{n^3\pi}[1 - (-1)^n]$$

$$f(x) = \frac{5\pi^2}{6} + \sum_{n=1}^{\infty}\left[\frac{2}{n^2}(-1)^{n+1}\cos nx + \left(\frac{\pi}{n}(-1)^n + \frac{2[1 - (-1^n]}{n^3\pi}\right)\sin nx\right]$$

9. $a_0 = \dfrac{1}{\pi}\displaystyle\int_{-\pi}^{\pi} f(x)\,dx = \dfrac{1}{\pi}\int_0^{\pi}\sin x\,dx = \dfrac{2}{\pi}$

$$a_n = \frac{1}{\pi}\int_{-\pi}^{\pi} f(x)\cos nx\,dx = \frac{1}{\pi}\int_0^{\pi}\sin x\,\cos nx\,dx = \frac{1}{2\pi}\int_0^{\pi}[\sin(n+1)x + \sin(1-n)x]\,dx$$

$$= \frac{1 + (-1)^n}{\pi(1 - n^2)} \quad \text{for } n = 2, 3, 4, \ldots$$

$$a_1 = \frac{1}{2\pi}\int_0^{\pi}\sin 2x\,dx = 0$$

$$b_n = \frac{1}{\pi}\int_{-\pi}^{\pi} f(x)\sin nx\,dx = \frac{1}{\pi}\int_0^{\pi}\sin x\,\sin nx\,dx$$

$$= \frac{1}{2\pi}\int_0^{\pi}[\cos(1-n)x - \cos(1+n)x]\,dx = 0 \quad \text{for } n = 2, 3, 4, \ldots$$

$$b_1 = \frac{1}{2\pi}\int_0^{\pi}(1 - \cos 2x)\,dx = \frac{1}{2}$$

$$f(x) = \frac{1}{\pi} + \frac{1}{2}\sin x + \sum_{n=2}^{\infty}\frac{1 + (-1)^n}{\pi(1 - n^2)}\cos nx$$

12. $a_0 = \dfrac{1}{2}\displaystyle\int_{-2}^{2} f(x)\,dx = \dfrac{1}{2}\left(\int_0^1 x\,dx + \int_1^2 1\,dx\right) = \dfrac{3}{4}$

$$a_n = \frac{1}{2}\int_{-2}^{2} f(x)\cos\frac{n\pi}{2}x\,dx = \frac{1}{2}\left(\int_0^1 x\cos\frac{n\pi}{2}x\,dx + \int_1^2\cos\frac{n\pi}{2}x\,dx\right) = \frac{2}{n^2\pi^2}\left(\cos\frac{n\pi}{2} - 1\right)$$

$$b_n = \frac{1}{2}\int_{-2}^{2} f(x)\sin\frac{n\pi}{2}x\,dx = \frac{1}{2}\left(\int_0^1 x\sin\frac{n\pi}{2}x\,dx + \int_1^2\sin\frac{n\pi}{2}x\,dx\right)$$

$$= \frac{2}{n^2\pi^2}\left(\sin\frac{n\pi}{2} + \frac{n\pi}{2}(-1)^{n+1}\right)$$

$$f(x) = \frac{3}{8} + \sum_{n=1}^{\infty}\left[\frac{2}{n^2\pi^2}\left(\cos\frac{n\pi}{2} - 1\right)\cos\frac{n\pi}{2}x + \frac{2}{n^2\pi^2}\left(\sin\frac{n\pi}{2} + \frac{n\pi}{2}(-1)^{n+1}\right)\sin\frac{n\pi}{2}x\right]$$

15. $a_0 = \frac{1}{\pi}\int_{-\pi}^{\pi} f(x)\,dx = \frac{1}{\pi}\int_{-\pi}^{\pi} e^x\,dx = \frac{1}{\pi}(e^\pi - e^{-\pi})$

$$a_n = \frac{1}{\pi}\int_{-\pi}^{\pi} f(x)\cos nx\,dx = \frac{(-1)^n(e^\pi - e^{-\pi})}{\pi(1+n^2)}$$

$$b_n = \frac{1}{\pi}\int_{-\pi}^{\pi} f(x)\sin nx\,dx = \frac{1}{\pi}\int_{-\pi}^{\pi} e^x\sin nx\,dx = \frac{(-1)^n n(e^{-\pi} - e^\pi)}{\pi(1+n^2)}$$

$$f(x) = \frac{e^\pi - e^{-\pi}}{2\pi} + \sum_{n=1}^{\infty}\left[\frac{(-1)^n(e^\pi - e^{-\pi})}{\pi(1+n^2)}\cos nx + \frac{(-1)^n n(e^{-\pi} - e^\pi)}{\pi(1+n^2)}\sin nx\right]$$

18. From Problem 17

$$\frac{\pi^2}{8} = \frac{1}{2}\left(\frac{\pi^2}{6} + \frac{\pi^2}{12}\right) = \frac{1}{2}\left(2 + \frac{2}{3^2} + \frac{2}{5^2} + \cdots\right) = 1 + \frac{1}{3^2} + \frac{1}{5^2} + \cdots.$$

21. **(a)** Letting $c_0 = a_0/2$, $c_n = (a_n - ib_n)$, and $c_{-n} = (a_n + ib_n)/2$ we have

$$\begin{aligned}
f(x) &= \frac{a_0}{2} + \sum_{n=1}^{\infty}\left(a_n\cos\frac{n\pi}{p}x + b_n\sin\frac{n\pi}{p}x\right)\\
&= c_0 + \sum_{n=1}^{\infty}\left(a_n\frac{e^{in\pi x/p} + e^{-in\pi x/p}}{2} + b_n\frac{e^{in\pi x/p} - e^{-in\pi x/p}}{2i}\right)\\
&= c_0 + \sum_{n=1}^{\infty}\left(a_n\frac{e^{in\pi x/p} + e^{-in\pi x/p}}{2} - b_n\frac{ie^{in\pi x/p} - ie^{-in\pi x/p}}{2}\right)\\
&= c_0 + \sum_{n=1}^{\infty}\left(\frac{a_n - ib_n}{2}e^{in\pi x/p} + \frac{a_n + ib_n}{2}e^{-in\pi x/p}\right)\\
&= c_0 + \sum_{n=1}^{\infty}\left(c_n e^{in\pi x/p} + c_{-n}e^{i(-n)\pi x/p}\right) = \sum_{n=-\infty}^{\infty} c_n e^{in\pi x/p}.
\end{aligned}$$

(b) Multiplying both sides of the expression in (a) by $e^{-im\pi x/p}$ and integrating we obtain

$$\begin{aligned}
\int_{-p}^{p} f(x)e^{-im\pi x/p}dx &= \int_{-p}^{p}\left(\sum_{n=-\infty}^{\infty} c_n e^{in\pi x/p}e^{-im\pi x/p}\right)dx\\
&= \sum_{n=-\infty}^{\infty} c_n\int_{-p}^{p} e^{i(n-m)\pi x/p}dx
\end{aligned}$$

$$= \sum_{n \neq m} c_n \int_{-p}^{p} e^{i(n-m)\pi x/p} dx + c_m \int_{-p}^{p} e^{i(m-n)\pi x/p} dx$$

$$= \sum_{n \neq m} c_n \int_{-p}^{p} e^{i(n-m)\pi x/p} dx + c_m \int_{-p}^{p} dx$$

$$= \sum_{n \neq m} c_n \int_{-p}^{p} e^{i(n-m)\pi x/p} dx + 2pc_m.$$

Recalling that

$$e^{iy} = \cos y + i \sin y \quad \text{and} \quad e^{-iy} = \cos y - i \sin y$$

we have for $n - m$ an integer and $n \neq m$

$$\begin{aligned}
\int_{-p}^{p} e^{i(n-m)\pi x/p} dx &= \frac{p}{i(n-m)\pi} e^{i(n-m)\pi x/p} \Big|_{-p}^{p} \\
&= \frac{p}{i(n-m)\pi} \left(e^{i(n-m)\pi} - e^{-i(n-m)\pi} \right) \\
&= \frac{p}{i(n-m)\pi} [\cos(n-m)\pi + i \sin(n-m)\pi - \cos(n-m)\pi + i \sin(n-m)\pi] \\
&= 0.
\end{aligned}$$

Thus

$$\int_{-p}^{p} f(x) e^{-im\pi x/p} dx = 2pc_m$$

and

$$c_m = \frac{1}{2p} \int_{-p}^{p} f(x) e^{-im\pi x/p} dx.$$

Exercises 11.3

3. Since

$$f(-x) = (-x)^2 - x = x^2 - x,$$

$f(x)$ is neither even nor odd.

6. Since

$$f(-x) = \left|(-x)^5\right| = \left|x^5\right| = f(x),$$

$f(x)$ is an even function.

9. Since $f(x)$ is not defined for $x < 0$, it is neither even nor odd.

12. Since $f(x)$ is an even function, we expand in a cosine series:

$$a_0 = \int_1^2 1\,dx = 1$$

$$a_n = \int_1^2 \cos\frac{n\pi}{2}x\,dx = -\frac{2}{n\pi}\sin\frac{n\pi}{2}.$$

Thus

$$f(x) = \frac{1}{2} + \sum_{n=1}^{\infty} \frac{-2}{n\pi}\sin\frac{n\pi}{2}\cos\frac{n\pi}{2}x.$$

15. Since $f(x)$ is an even function, we expand in a cosine series:

$$a_0 = 2\int_0^1 x^2\,dx = \frac{2}{3}$$

$$a_n = 2\int_0^1 x^2\cos n\pi x\,dx = 2\left(\frac{x^2}{n\pi}\sin n\pi x\,\Big|_0^1 - \frac{2}{n\pi}\int_0^1 x\sin n\pi x\,dx\right) = \frac{4}{n^2\pi^2}(-1)^n.$$

Thus

$$f(x) = \frac{1}{3} + \sum_{n=1}^{\infty} \frac{4}{n^2\pi^2}(-1)^n\cos n\pi x.$$

18. Since $f(x)$ is an odd function, we expand in a sine series:

$$b_n = \frac{2}{\pi}\int_0^\pi x^3\sin nx\,dx = \frac{2}{\pi}\left(-\frac{x^3}{n}\cos nx\,\Big|_0^\pi + \frac{3}{n}\int_0^\pi x^2\cos nx\,dx\right)$$

$$= \frac{2\pi^2}{n}(-1)^{n+1} - \frac{12}{n^2\pi}\int_0^\pi x\sin nx\,dx$$

$$= \frac{2\pi^2}{n}(-1)^{n+1} - \frac{12}{n^2\pi}\left(-\frac{x}{n}\cos nx\,\Big|_0^\pi + \frac{1}{n}\int_0^\pi \cos nx\,dx\right) = \frac{2\pi^2}{n}(-1)^{n+1} + \frac{12}{n^3}(-1)^n.$$

Thus

$$f(x) = \sum_{n=1}^{\infty}\left(\frac{2\pi^2}{n}(-1)^{n+1} + \frac{12}{n^3}(-1)^n\right)\sin nx.$$

21. Since $f(x)$ is an even function, we expand in a cosine series:

$$a_0 = \int_0^1 x\,dx + \int_1^2 1\,dx = \frac{3}{2}$$

$$a_n = \int_0^1 x\cos\frac{n\pi}{2}x\,dx + \int_1^2 \cos\frac{n\pi}{2}x\,dx = \frac{4}{n^2\pi^2}\left(\cos\frac{n\pi}{2} - 1\right).$$

Thus

$$f(x) = \frac{3}{4} + \sum_{n=1}^{\infty}\frac{4}{n^2\pi^2}\left(\cos\frac{n\pi}{2} - 1\right)\cos\frac{n\pi}{2}x.$$

24. Since $f(x)$ is an even function, we expand in a cosine series:

$$a_0 = \frac{2}{\pi/2}\int_0^{\pi/2} \cos x\,dx = \frac{4}{\pi}$$

$$a_n = \frac{2}{\pi/2}\int_0^{\pi/2} \cos x \cos\frac{n\pi}{\pi/2}x\,dx = \frac{4}{\pi}\int_0^{\pi/2} \cos x \cos 2nx\,dx$$

$$= \frac{2}{\pi}\int_0^{\pi/2} [\cos(2n-1)x + \cos(2n+1)x]\,dx = \frac{4(-1)^{n+1}}{\pi\,(4n^2-1)}.$$

Thus

$$f(x) = \frac{2}{\pi} + \sum_{n=1}^{\infty} \frac{4(-1)^{n+1}}{\pi\,(4n^2-1)} \cos 2nx.$$

27. $a_0 = \dfrac{4}{\pi}\displaystyle\int_0^{\pi/2} \cos x\,dx = \dfrac{4}{\pi}$

$$a_n = \frac{4}{\pi}\int_0^{\pi/2} \cos x\,\cos 2nx\,dx = \frac{2}{\pi}\int_0^{\pi/2} [\cos(2n+1)x + \cos(2n-1)x]\,dx = \frac{4(-1)^n}{\pi(1-4n^2)}$$

$$b_n = \frac{4}{\pi}\int_0^{\pi/2} \cos x\,\sin 2nx\,dx = \frac{2}{\pi}\int_0^{\pi/2} [\sin(2n+1)x + \sin(2n-1)x]\,dx = \frac{8n}{\pi(4n^2-1)}$$

$$f(x) = \frac{2}{\pi} + \sum_{n=1}^{\infty} \frac{4(-1)^n}{\pi(1-4n^2)} \cos 2nx$$

$$f(x) = \sum_{n=1}^{\infty} \frac{8n}{\pi(4n^2-1)} \sin 2nx$$

30. $a_0 = \dfrac{1}{\pi}\displaystyle\int_\pi^{2\pi} (x-\pi)\,dx = \dfrac{\pi}{2}$

$$a_n = \frac{1}{\pi}\int_\pi^{2\pi} (x-\pi)\cos\frac{n}{2}x\,dx = \frac{4}{n^2\pi}\left[(-1)^n - \cos\frac{n\pi}{2}\right]$$

$$b_n = \frac{1}{\pi}\int_\pi^{2\pi} (x-\pi)\sin\frac{n}{2}x\,dx = \frac{2}{n}(-1)^{n+1} - \frac{4}{n^2\pi}\sin\frac{n\pi}{2}$$

$$f(x) = \frac{\pi}{4} + \sum_{n=1}^{\infty} \frac{4}{n^2\pi}\left[(-1)^n - \cos\frac{n\pi}{2}\right]\cos\frac{n}{2}x$$

$$f(x) = \sum_{n=1}^{\infty} \left(\frac{2}{n}(-1)^{n+1} - \frac{4}{n^2\pi}\sin\frac{n\pi}{2}\right)\sin\frac{n}{2}x$$

33. $a_0 = 2\displaystyle\int_0^1 (x^2+x)\,dx = \dfrac{5}{3}$

$$a_n = 2\int_0^1 (x^2+x)\cos n\pi x\,dx = \frac{2(x^2+x)}{n\pi}\sin n\pi x\,\Big|_0^1 - \frac{2}{n\pi}\int_0^1 (2x+1)\sin n\pi x\,dx = \frac{2}{n^2\pi^2}[3(-1)^n-1]$$

$$b_n = 2\int_0^1 (x^2+x)\sin n\pi x\,dx = -\frac{2(x^2+x)}{n\pi}\cos n\pi x\,\Big|_0^1 + \frac{2}{n\pi}\int_0^1 (2x+1)\cos n\pi x\,dx$$

$$= \frac{4}{n\pi}(-1)^{n+1} + \frac{4}{n^3\pi^3}[(-1)^n-1]$$

$$f(x) = \frac{5}{6} + \sum_{n=1}^{\infty} \frac{2}{n^2\pi^2}[3(-1)^n-1]\cos n\pi x$$

$$f(x) = \sum_{n=1}^{\infty}\left(\frac{4}{n\pi}(-1)^{n+1} + \frac{4}{n^3\pi^3}[(-1)^n-1]\right)\sin n\pi x$$

36. $a_0 = \dfrac{2}{\pi}\displaystyle\int_0^{\pi} x\,dx = \pi$

$$a_n = \frac{2}{\pi}\int_0^{\pi} x\cos 2nx\,dx = 0$$

$$b_n = \frac{2}{\pi}\int_0^{\pi} x\sin 2nx\,dx = -\frac{1}{n}$$

$$f(x) = \frac{\pi}{2} + \sum_{n=1}^{\infty}\left(-\frac{1}{n}\sin 2nx\right)$$

39. We have

$$b_n = \frac{2}{\pi}\int_0^{\pi} 5\sin nt\,dt = \frac{10}{n\pi}[1-(-1)^n]$$

so that

$$f(t) = \sum_{n=1}^{\infty}\frac{10[1-(-1)^n]}{n\pi}\sin nt.$$

Substituting the assumption $x_p(t) = \sum_{n=1}^{\infty} B_n \sin nt$ into the differential equation then gives

$$x_p'' + 10x_p = \sum_{n=1}^{\infty} B_n(10-n^2)\sin nt = \sum_{n=1}^{\infty}\frac{10[1-(-1)^n]}{n\pi}\sin nt$$

and so $B_n = \dfrac{10[1-(-1)^n]}{n\pi(10-n^2)}$. Thus

$$x_p(t) = \frac{10}{\pi}\sum_{n=1}^{\infty}\frac{1-(-1)^n}{n(10-n^2)}\sin nt.$$

42. We have

$$a_0 = \frac{2}{(1/2)} \int_0^{1/2} t\, dt = \frac{1}{2}$$

$$a_n = \frac{2}{(1/2)} \int_0^{1/2} t \cos 2n\pi t\, dt = \frac{1}{n^2\pi^2} [(-1)^n - 1]$$

so that

$$f(t) = \frac{1}{4} + \sum_{n=1}^{\infty} \frac{(-1)^n - 1}{n^2\pi^2} \cos 2n\pi t.$$

Substituting the assumption

$$x_p(t) = \frac{A_0}{2} + \sum_{n=1}^{\infty} A_n \cos 2n\pi t$$

into the differential equation then gives

$$\frac{1}{4} x_p'' + 12x_p = 6A_0 + \sum_{n=1}^{\infty} A_n(12 - n^2\pi^2) \cos 2n\pi t = \frac{1}{4} + \sum_{n=1}^{\infty} \frac{(-1)^n - 1}{n^2\pi^2} \cos 2n\pi t$$

and $A_0 = \dfrac{1}{24}$, $A_n = \dfrac{(-1)^n - 1}{n^2\pi^2(12 - n^2\pi^2)}$. Thus

$$x_p(t) = \frac{1}{48} + \frac{1}{\pi^2} \sum_{n=1}^{\infty} \frac{(-1)^n - 1}{n^2(12 - n^2\pi^2)} \cos 2n\pi t.$$

45. If f and g are even and $h(x) = f(x)g(x)$ then

$$h(-x) = f(-x)g(-x) = f(x)g(x) = h(x)$$

and h is even.

48. If f is even then

$$\int_{-a}^{a} f(x)\, dx = -\int_a^0 f(-u)\, du + \int_0^a f(x)\, dx = \int_0^a f(u)\, du + \int_0^a f(x)\, dx = 2\int_0^a f(x)\, dx.$$

51. Using Problems 13 and 14 we obtain

$$f(x) = \frac{1}{2}|x| + \frac{1}{2}x = \frac{\pi}{4} + \sum_{n=1}^{\infty} \left(\frac{(-1)^n - 1}{\pi n^2} \cos nx + \frac{(-1)^{n+1}}{n} \sin nx \right).$$

Exercises 11.4

3. For $\lambda = 0$ the solution of $y'' = 0$ is $y = c_1x + c_2$. The condition $y'(0) = 0$ implies $c_1 = 0$, so $\lambda = 0$ is an eigenvalue with corresponding eigenfunction 1. For $\lambda < 0$ we have $y = c_1 \cosh\sqrt{-\lambda}\,x + c_2 \sin\sqrt{-\lambda}\,x$ and $y' = c_1\sqrt{-\lambda}\sinh\sqrt{-\lambda}\,x + c_2\sqrt{-\lambda}\cosh\sqrt{-\lambda}\,x$. The condition $y'(0) = 0$ implies $c_2 = 0$ and so $y = c_1 \cosh\sqrt{-\lambda}\,x$. Now the condition $y'(L) = 0$ implies $c_1 = 0$. Thus $y = 0$ and there are no negative eigenvalues. For $\lambda > 0$ we have $y = c_1 \cos\sqrt{\lambda}\,x + c_2 \sin\sqrt{\lambda}\,x$ and $y' = -c_1\sqrt{\lambda}\sin\sqrt{\lambda}\,x + c_2\sqrt{\lambda}\cos\sqrt{\lambda}\,x$. The condition $y'(0) = 0$ implies $c_2 = 0$ and so $y = c_1 \cos\sqrt{\lambda}\,x$. Now the condition $y'(L) = 0$ implies $-c_1\sqrt{\lambda}\sin\sqrt{\lambda}\,L = 0$. For $c_1 \neq 0$ this condition will hold when $\sqrt{\lambda}\,L = n\pi$ or $\lambda = n^2\pi^2/L^2$, where $n = 1, 2, 3, \ldots$. These are the positive eigenvalues with corresponding eigenfunctions $\cos(n\pi/L)x$, $n = 1, 2, 3, \ldots$.

6. The eigenfunctions are $\sin\sqrt{\lambda_n}\,x$ where $\tan\sqrt{\lambda_n} = -\lambda_n$. Thus

$$\begin{aligned}
\|\sin\sqrt{\lambda_n}\,x\|^2 &= \int_0^1 \sin^2\sqrt{\lambda_n}\,x\,dx = \frac{1}{2}\int_0^1\left(1 - \cos 2\sqrt{\lambda_n}\,x\right)dx \\
&= \frac{1}{2}\left(x - \frac{1}{2\sqrt{\lambda_n}}\sin 2\sqrt{\lambda_n}\,x\right)\Big|_0^1 = \frac{1}{2}\left(1 - \frac{1}{2\sqrt{\lambda_n}}\sin 2\sqrt{\lambda_n}\right) \\
&= \frac{1}{2}\left[1 - \frac{1}{2\sqrt{\lambda_n}}\left(2\sin\sqrt{\lambda_n}\cos\sqrt{\lambda_n}\right)\right] \\
&= \frac{1}{2}\left[1 - \frac{1}{\sqrt{\lambda_n}}\tan\sqrt{\lambda_n}\cos\sqrt{\lambda_n}\cos\sqrt{\lambda_n}\right] \\
&= \frac{1}{2}\left[1 - \frac{1}{\sqrt{\lambda_n}}\left(-\sqrt{\lambda_n}\cos^2\sqrt{\lambda_n}\right)\right] = \frac{1}{2}\left(1 + \cos^2\sqrt{\lambda_n}\right).
\end{aligned}$$

9. (a) An orthogonality relation is

$$\int_0^1 \cos x_m x \cos x_n x = 0$$

where $x_m \neq x_n$ are positive solutions of $\cot x = x$.

(b) Referring to Problem 1 we use a CAS to compute

$$\int_0^1 (\cos 0.8603x)(\cos 3.4256x)\,dx = -1.8771 \times 10^{-6}.$$

12. To obtain the self-adjoint form we note that an integrating factor is $e^{\int -2x\,dx} = e^{-x^2}$. Thus, the differential equation is

$$e^{-x^2}y'' - 2xe^{-x^2}y' + 2ne^{-x^2}y = 0$$

and the self-adjoint form is

$$\frac{d}{dx}\left[e^{-x^2}y'\right] + 2ne^{-x^2}y = 0.$$

Identifying the weight function $p(x) = 2e^{-x^2}$ and noting that since $r(x) = e^{-x^2}$, $\lim_{x\to-\infty} r(x) = \lim_{x\to\infty} r(x) = 0$, we have the orthogonality relation

$$\int_{\infty}^{\infty} 2e^{-x^2} H_n(x)H_m(x)\,dx = 0, \; m \neq n.$$

Exercises 11.5

3. The boundary condition indicates that we use (15) and (16) of Section 11.5. With $b = 2$ we obtain

$$c_i = \frac{2}{4J_1^2(2\lambda_i)} \int_0^2 xJ_0(\lambda_i x)\,dx$$

$t = \lambda_i x$	$dt = \lambda_i\,dx$

$$= \frac{1}{2J_1^2(2\lambda_i)} \cdot \frac{1}{\lambda_i^2} \int_0^{2\lambda_i} tJ_0(t)\,dt$$

$$= \frac{1}{2\lambda_i^2 J_1^2(2\lambda_i)} \int_0^{2\lambda_i} \frac{d}{dt}[tJ_1(t)]\,dt \qquad \text{[From (4) in the text]}$$

$$= \frac{1}{2\lambda_i^2 J_1^2(2\lambda_i)} tJ_1(t)\Big|_0^{2\lambda_i}$$

$$= \frac{1}{\lambda_i J_1(2\lambda_i)}.$$

Thus

$$f(x) = \sum_{i=1}^{\infty} \frac{1}{\lambda_i J_1(2\lambda_i)} J_0(\lambda_i x).$$

6. Writing the boundary condition in the form

$$2J_0(2\lambda) + 2\lambda J_0'(2\lambda) = 0$$

we identify $b = 2$ and $h = 2$. Using (17) and (18) of Section 11.5 we obtain

$$c_i = \frac{2\lambda_i^2}{(4\lambda_i^2 + 4)J_0^2(2\lambda_i)} \int_0^2 xJ_0(\lambda_i x)\,dx$$

$t = \lambda_i x$	$dt = \lambda_i\,dx$

$$= \frac{\lambda_i^2}{2(\lambda_i^2+1)J_0^2(2\lambda_i)} \cdot \frac{1}{\lambda_i^2} \int_0^{2\lambda_i} tJ_0(t)\,dt$$

$$= \frac{1}{2(\lambda_i^2+1)J_0^2(2\lambda_i)} \int_0^{2\lambda_i} \frac{d}{dt}[tJ_1(t)]\,dt \qquad \text{[From (4) in the text]}$$

$$= \frac{1}{2(\lambda_i^2+1)J_0^2(2\lambda_i)} tJ_1(t)\Big|_0^{2\lambda_i}$$

$$= \frac{\lambda_i J_1(2\lambda_i)}{(\lambda_i^2+1)J_0^2(2\lambda_i)}.$$

Thus

$$f(x) = \sum_{i=1}^{\infty} \frac{\lambda_i J_1(2\lambda_i)}{(\lambda_i^2+1)J_0^2(2\lambda_i)} J_0(\lambda_i x).$$

9. The boundary condition indicates that we use (19) and (20) of Section 11.5. With $b=3$ we obtain

$$c_1 = \frac{2}{9}\int_0^3 x x^2\,dx = \frac{2}{9}\frac{x^4}{4}\Big|_0^3 = \frac{9}{2},$$

$$c_i = \frac{2}{9J_0^2(3\lambda_i)} \int_0^3 xJ_0(\lambda_i x)x^2\,dx$$

$t=\lambda_i x$	$dt=\lambda_i\,dx$

$$= \frac{2}{9J_0^2(3\lambda_i)} \cdot \frac{1}{\lambda_i^4} \int_0^{3\lambda_i} t^3 J_0(t)\,dt$$

$$= \frac{2}{9\lambda_i^4 J_0^2(3\lambda_i)} \int_0^{3\lambda_i} t^2 \frac{d}{dt}[tJ_1(t)]\,dt$$

$u=t^2$	$dv=\frac{d}{dt}[tJ_1(t)]\,dt$
$du=2t\,dt$	$v=tJ_1(t)$

$$= \frac{2}{9\lambda_i^4 J_0^2(3\lambda_i)} \left(t^3 J_1(t)\Big|_0^{3\lambda_i} - 2\int_0^{3\lambda_i} t^2 J_1(t)\,dt \right)$$

With $n=0$ in equation (5) in Section 11.5 in the text we have $J_0'(x) = -J_1(x)$, so the boundary condition $J_0'(3\lambda_i) = 0$ implies $J_1(3\lambda_i) = 0$. Then

$$c_i = \frac{2}{9\lambda_i^4 J_0^2(3\lambda_i)}\left(-2\int_0^{3\lambda_i}\frac{d}{dt}\left[t^2 J_2(t)\right]dt\right) = \frac{2}{9\lambda_i^4 J_0^2(3\lambda_i)}\left(-2t^2 J_2(t)\Big|_0^{3\lambda_i}\right)$$

$$= \frac{2}{9\lambda_i^4 J_0^2(3\lambda_i)}\left[-18\lambda_i^2 J_2(3\lambda_i)\right] = \frac{-4J_2(3\lambda_i)}{\lambda_i^2 J_0^2(3\lambda_i)}.$$

Thus

$$f(x) = \frac{9}{2} - 4\sum_{i=1}^{\infty}\frac{J_2(3\lambda_i)}{\lambda_i^2 J_0^2(3\lambda_i)}J_0(\lambda_i x).$$

12. Since $f(x) = x^3$ is a polynomial in x, an expansion of f in polynomials in x must terminate with the term having the same degree as f. We have

$$c_0 = \frac{1}{2}\int_{-1}^{1} x^3 P_0(x)\,dx = \frac{1}{2}\int_{-1}^{1} x^3\,dx = 0,$$

$$c_1 = \frac{3}{2}\int_{-1}^{1} x^3 P_1(x)\,dx = \frac{3}{2}\int_{-1}^{1} x^4\,dx = \frac{3}{5},$$

$$c_2 = \frac{5}{2}\int_{-1}^{1} x^3 P_2(x)\,dx = \frac{5}{2}\int_{-1}^{1} x^3\frac{1}{2}\left(3x^2-1\right)dx = 0,$$

$$c_3 = \frac{7}{2}\int_{-1}^{1} x^3 P_3(x)\,dx = \frac{7}{2}\int_{-1}^{1} x^3\frac{1}{2}\left(5x^3-3x\right)dx = \frac{2}{5}.$$

Thus

$$f(x) = c_0P_0(x) + c_1P_1(x) + c_2P_2(x) + c_3P_3(x)$$

$$= \frac{3}{5}P_1(x) + \frac{2}{5}P_3(x).$$

15. Using $\cos^2\theta = \frac{1}{2}(\cos 2\theta + 1)$ we have

$$P_2(\cos\theta) = \frac{1}{2}(3\cos^2\theta - 1) = \frac{3}{2}\cos^2\theta - \frac{1}{2}$$

$$= \frac{3}{4}(\cos 2\theta + 1) - \frac{1}{2} = \frac{3}{4}\cos 2\theta + \frac{1}{4} = \frac{1}{4}(3\cos 2\theta + 1).$$

18. If f is an odd function on $(-1, 1)$ then

$$\int_{-1}^{1} f(x)P_{2n}(x)\,dx = 0$$

and

$$\int_{-1}^{1} f(x)P_{2n+1}(x)\,dx = 2\int_0^1 f(x)P_{2n+1}(x)\,dx.$$

Thus

$$c_{2n+1} = \frac{2(2n+1)+1}{2}\int_{-1}^{1} f(x)P_{2n+1}(x)\,dx = \frac{4n+3}{2}\left(2\int_0^1 f(x)P_{2n+1}(x)\,dx\right)$$

$$= (4n+1)\int_0^1 f(x)P_{2n+1}(x)\,dx,$$

$c_{2n} = 0$, and

$$f(x) = \sum_{n=0}^{\infty} c_{2n+1}P_{2n+1}(x).$$

Chapter 11 Review Exercises

3. cosine, since f is even.

6. True

9. Since the coefficient of y in the differential equation is n^2, the weight function is the integrating factor

$$\frac{1}{a(x)}e^{\int (b/a)dx} = \frac{1}{1-x^2}e^{\int -\frac{x}{1-x^2}\,dx} = \frac{1}{1-x^2}e^{\frac{1}{2}\ln(1-x^2)} = \frac{\sqrt{1-x^2}}{1-x^2} = \frac{1}{\sqrt{1-x^2}}$$

on the interval $[-1, 1]$.

12. From

$$\int_0^L \sin^2\frac{(2n+1)\pi}{2L}x\,dx = \int_0^L\left(\frac{1}{2}-\frac{1}{2}\cos\frac{(2n+1)\pi}{2L}x\right)dx = \frac{L}{2}$$

we see that

$$\left\|\sin\frac{(2n+1)\pi}{2L}x\right\| = \sqrt{\frac{L}{2}}.$$

15. Since

$$A_0 = 2\int_0^1 e^{-x}dx$$

and

$$A_n = 2\int_{-1}^{1} e^{-x}\cos n\pi x\,dx = \frac{2}{1+n^2\pi^2}[(1-(-1)^n e^{-1}]$$

for $n = 1, 2, 3, \ldots$ we have

$$f(x) = 1 - e^{-1} + 2\sum_{n=1}^{\infty}\frac{1-(-1)^n e^{-1}}{1+n^2\pi^2}\cos n\pi x.$$

18. To obtain the self-adjoint form of the differential equation in Problem 17 we note that an integrating factor is $(1/x^2)e^{\int dx/x} = 1/x$. Thus the weight function is $9/x$ and an orthogonality relation is

$$\int_1^e \frac{9}{x}\cos\left(\frac{2n-1}{2}\pi\ln x\right)\cos\left(\frac{2m-1}{2}\pi\ln x\right)dx = 0,\ m \neq n.$$

12 Partial Differential Equations and Boundary-Value Problems in Rectangular Coordinates

Exercises 12.1

3. If $u = XY$ then

$$u_x = X'Y,$$
$$u_y = XY',$$
$$X'Y = X(Y - Y'),$$

and

$$\frac{X'}{X} = \frac{Y - Y'}{Y} = \pm\lambda^2.$$

Then

$$X' \mp \lambda^2 X = 0 \quad \text{and} \quad Y' - (1 \mp \lambda^2)Y = 0$$

so that

$$X = A_1 e^{\pm\lambda^2 x},$$
$$Y = A_2 e^{(1\mp\lambda^2)y},$$

and

$$u = XY = c_1 e^{y + c_2(x-y)}.$$

6. If $u = XY$ then

$$u_x = X'Y,$$
$$u_y = XY',$$
$$yX'Y = xXY',$$

and

$$\frac{X'}{xX} = \frac{Y'}{-yY} = \pm\lambda^2.$$

Then

$$X \mp \lambda^2 xX = 0 \quad \text{and} \quad Y' \pm \lambda^2 yY = 0$$

so that

$$X = A_1 e^{\pm\lambda^2 x^2/2},$$

$$Y = A_2 e^{\mp\lambda^2 y^2/2},$$

and

$$u = XY = c_1 e^{c_2(x^2-y^2)}.$$

9. If $u = XT$ then

$$u_t = XT',$$

$$u_{xx} = X''T,$$

$$kX''T - XT = XT',$$

and we choose

$$\frac{T'}{T} = \frac{kX'' - X}{X} = -1 \pm k\lambda^2$$

so that

$$T' - (-1 \pm k\lambda^2)T = 0 \quad \text{and} \quad X'' - (\pm\lambda^2)X = 0.$$

For $\lambda^2 > 0$ we obtain

$$X = A_1 \cosh \lambda x + A_2 \sinh \lambda x \quad \text{and} \quad T = A_3 e^{(-1+k\lambda^2)t}$$

so that

$$u = XT = e^{(-1+k\lambda^2)t}\left(c_1 \cosh \lambda x + c_2 \sinh \lambda x\right).$$

For $-\lambda^2 < 0$ we obtain

$$X = A_1 \cos \lambda x + A_2 \sin \lambda x \quad \text{and} \quad T = A_3 e^{(-1-k\lambda^2)t}$$

so that

$$u = XT = e^{(-1-k\lambda^2)t}(c_3 \cos \lambda x + c_4 \sin \lambda x).$$

If $\lambda^2 = 0$ then

$$X'' = 0 \quad \text{and} \quad T' + T = 0,$$

and we obtain

$$X = A_1 x + A_2 \quad \text{and} \quad T = A_3 e^{-t}.$$

In this case

$$u = XT = e^{-t}(c_5 x + c_6)$$

12. If $u = XT$ then

$$u_t = XT',$$

$$u_{tt} = XT'',$$

$$u_{xx} = X''T,$$

$$a^2X''T = XT'' + 2kXT',$$

and

$$\frac{X''}{X} = \frac{T'' + 2kT'}{a^2T} = \pm\lambda^2$$

so that

$$X'' \mp \lambda^2 X = 0 \quad \text{and} \quad T'' + 2kT' \mp a^2\lambda^2 T = 0.$$

For $\lambda^2 > 0$ we obtain

$$X = A_1 e^{\lambda x} + A_2 e^{-\lambda x},$$

$$T = A_3 e^{(-k+\sqrt{k^2+a^2\lambda^2})t} + A_4 e^{(-k-\sqrt{k^2+a^2\lambda^2})t},$$

and

$$u = XT = \left(A_1 e^{\lambda x} + A_2 e^{-\lambda x}\right)\left(A_3 e^{(-k+\sqrt{k^2+a^2\lambda^2})t} + A_4 e^{(-k-\sqrt{k^2+a^2\lambda^2})t}\right).$$

For $-\lambda^2 < 0$ we obtain

$$X = A_1 \cos\lambda x + A_2 \sin\lambda x.$$

If $k^2 - a^2\lambda^2 > 0$ then

$$T = A_3 e^{(-k+\sqrt{k^2-a^2\lambda^2})t} + A_4 e^{(-k-\sqrt{k^2-a^2\lambda^2})t}.$$

If $k^2 - a^2\lambda^2 < 0$ then

$$T = e^{-kt}\left(A_3 \cos\sqrt{a^2\lambda^2 - k^2}\,t + A_4 \sin\sqrt{a^2\lambda^2 - k^2}\,t\right).$$

If $k^2 - a^2\lambda^2 = 0$ then

$$T = A_3 e^{-kt} + A_4 t e^{-kt}$$

so that

$$u = XT = (A_1 \cos\lambda x + A_2 \sin\lambda x)\left(A_3 e^{(-k+\sqrt{k^2-a^2\lambda^2})t} + A_4 e^{(-k-\sqrt{k^2-a^2\lambda^2})t}\right)$$

$$= (A_1 \cos\lambda x + A_2 \sin\lambda x)e^{-kt}\left(A_3 \cos\sqrt{a^2\lambda^2 - k^2}\,t + A_4 \sin\sqrt{a^2\lambda^2 - k^2}\,t\right)$$

$$= \left(A_1 \cos\frac{k}{a}x + A_2 \sin\frac{k}{a}x\right)\left(A_3 e^{-kt} + A_4 t e^{-kt}\right).$$

For $\lambda^2 = 0$ we obtain

$$XA_1 x + A_2,$$

$$T = A_3 + A_4 e^{-2kt},$$

and

$$u = XT = (A_1x + A_2)(A_3 + A_4e^{-2kt}).$$

15. If $u = XY$ then

$$u_{xx} = X''Y,$$

$$u_{yy} = XY'',$$

$$X''Y + XY'' = XY,$$

and

$$\frac{X''}{X} = \frac{Y - Y''}{Y} = \pm\lambda^2$$

so that

$$X'' \mp \lambda^2 X = 0 \quad \text{and} \quad Y'' + (\pm\lambda^2 - 1)Y = 0.$$

For $\lambda^2 > 0$ we obtain

$$X = A_1e^{\lambda x} + A_2e^{-\lambda x}.$$

If $\lambda^2 - 1 > 0$ then

$$Y = A_3 \cos\sqrt{\lambda^2 - 1}\,y + A_4 \sin\sqrt{\lambda^2 - 1}\,y.$$

If $\lambda^2 - 1 < 0$ then

$$Y = A_3e^{\sqrt{1-\lambda^2}\,y} + A_4e^{-\sqrt{1-\lambda^2}\,y}.$$

If $\lambda^2 - 1 = 0$ then $Y = A_3y + A_4$ so that

$$u = XY = \left(A_1e^{\lambda x} + A_2e^{-\lambda x}\right)\left(A_3 \cos\sqrt{\lambda^2 - 1}\,y + A_4 \sin\sqrt{\lambda^2 - 1}\,y\right),$$

$$= \left(A_1e^{\lambda x} + A_2e^{-\lambda x}\right)\left(A_3e^{\sqrt{1-\lambda^2}\,y} + A_4e^{-\sqrt{1-\lambda^2}\,y}\right)$$

$$= (A_1e^{x} + A_2e^{-x})(A_3y + A_4).$$

For $-\lambda^2 < 0$ we obtain

$$X = A_1 \cos\lambda x + A_2 \sin\lambda x,$$

$$Y = A_3e^{\sqrt{1+\lambda^2}\,y} + A_4e^{-\sqrt{1+\lambda^2}\,y},$$

and

$$u = XY = (A_1 \cos\lambda x + A_2 \sin\lambda x)\left(A_3e^{\sqrt{1+\lambda^2}\,y} + A_4e^{-\sqrt{1+\lambda^2}\,y}\right).$$

For $\lambda^2 = 0$ we obtain

$$X = A_1x + A_2,$$

$$Y = A_3e^{y} + A_4e^{-y},$$

and

$$u = XY = (A_1x + A_2)(A_3e^y + A_4e^{-y}).$$

18. Identifying $A = 3$, $B = 5$, and $C = 1$, we compute $B^2 - 4AC = 13 > 0$. The equation is hyperbolic.

21. Identifying $A = 1$, $B = -9$, and $C = 0$, we compute $B^2 - 4AC = 81 > 0$. The equation is hyperbolic.

24. Identifying $A = 1$, $B = 0$, and $C = 1$, we compute $B^2 - 4AC = -4 < 0$. The equation is elliptic.

27. If $u = RT$ then

$$u_r = R'T,$$
$$u_{rr} = R''T,$$
$$u_t = RT',$$
$$RT' = k\left(R''T + \frac{1}{r}R'T\right),$$

and

$$\frac{r^2R'' + rR'}{r^2R} = \frac{T'}{kT} = \pm\lambda^2.$$

If we use $-\lambda^2 < 0$ then

$$r^2R'' + rR' + \lambda^2r^2R = 0 \quad \text{and} \quad T'' \mp \lambda^2kT = 0$$

so that

$$T = A_1e^{-k\lambda^2t},$$
$$R = A_2J_0(\lambda r) + A_3Y_0(\lambda r),$$

and

$$u = RT = e^{-k\lambda^2t}[c_1J_0(\lambda r) + c_2Y_0(\lambda r)]$$

30. For $u = A_1e^{\lambda^2y}\cosh 2\lambda x + B_1e^{\lambda^2y}\sinh 2\lambda x$ we compute

$$\frac{\partial^2u}{\partial x^2} = 4\lambda^2A_1e^{\lambda^2y}\cosh 2\lambda x + 4\lambda^2B_1e^{\lambda^2y}\sinh 2\lambda x$$

and

$$\frac{\partial u}{\partial y} = \lambda^2A_1e^{\lambda^2y}\cosh 2\lambda x + \lambda^2B_1e^{\lambda^2y}\sinh 2\lambda x.$$

Then $\partial^2u/\partial x^2 = 4\partial u/\partial y$.

For $u = A_2e^{-\lambda^2y}\cos 2\lambda x + B_2e^{-\lambda^2y}\sin 2\lambda x$ we compute

$$\frac{\partial^2u}{\partial x^2} = -4\lambda^2A_2e^{-\lambda^2y}\cos 2\lambda x - 4\lambda^2B_2e^{-\lambda^2y}\sin 2\lambda x$$

and

$$\frac{\partial u}{\partial y} = -\lambda^2A_2\cos 2\lambda x - \lambda^2B_2\sin 2\lambda x.$$

Then $\partial^2 u/\partial x^2 = 4\partial u/\partial y$.

For $u = A_3 x + B_3$ we compute $\partial^2 u/\partial x^2 = \partial u/\partial y = 0$. Then $\partial^2 u/\partial x^2 = 4\partial u/\partial y$.

Exercises 12.2

3. $k\dfrac{\partial^2 u}{\partial x^2} = \dfrac{\partial u}{\partial t}, \quad 0 < x < L, \ t > 0$

$u(0,t) = 100, \quad \left.\dfrac{\partial u}{\partial x}\right|_{x=L} = -hu(L,t), \quad t > 0$

$u(x,0) = f(x), \quad 0 < x < L$

6. $a^2\dfrac{\partial^2 u}{\partial x^2} = \dfrac{\partial^2 u}{\partial t^2}, \quad 0 < x < L, \ t > 0$

$u(0,t) = 0, \quad u(L,t) = 0, \quad t > 0$

$u(x,0) = 0, \quad \left.\dfrac{\partial u}{\partial x}\right|_{t=0} = \sin\dfrac{\pi x}{L}, \quad 0 < x < L$

9. $\dfrac{\partial^2 u}{\partial x^2} + \dfrac{\partial^2 u}{\partial y^2} = 0, \quad 0 < x < 4, \ 0 < y < 2$

$\left.\dfrac{\partial u}{\partial x}\right|_{x=0} = 0, \quad u(4,y) = f(y), \quad 0 < y < 2$

$\left.\dfrac{\partial u}{\partial y}\right|_{y=0} = 0, \quad u(x,2) = 0, \quad 0 < x < 4$

Exercises 12.3

3. Using $u = XT$ and $-\lambda^2$ as a separation constant leads to

$$X'' + \lambda^2 X = 0,$$
$$X'(0) = 0,$$
$$X'(L) = 0,$$

and

$$T' + k\lambda^2 T = 0.$$

Then

$$X = c_1 \cos\frac{n\pi}{L}x \quad \text{and} \quad T = c_2 e^{-\frac{kn^2\pi^2}{L^2}t}$$

for $n = 0, 1, 2, \ldots$ so that

$$u = \sum_{n=0}^{\infty} A_n e^{-\frac{kn^2\pi^2}{L^2}t} \cos\frac{n\pi}{L}x.$$

Imposing

$$u(x,0) = f(x) = \sum_{n=0}^{\infty} A_n \cos\frac{n\pi}{L}x.$$

gives

$$u(x,t) = \frac{1}{L}\int_0^L f(x)\,dx + \frac{2}{L}\sum_{n=1}^{\infty}\left(\int_0^L f(x)\cos\frac{n\pi}{L}x\,dx\right) e^{-\frac{kn^2\pi^2}{L^2}t}\cos\frac{n\pi}{L}x.$$

6. Using $u = XT$ and $-\lambda^2$ as a separation constant leads to

$$X'' + \lambda^2 X = 0,$$

$$X(0) = 0,$$

$$X(L) = 0,$$

and

$$T' + (h + k\lambda^2)T = 0.$$

Then

$$X = c_1 \sin\frac{n\pi}{L}x \quad \text{and} \quad T = c_2 e^{-\left(h+\frac{kn^2\pi^2}{L^2}\right)t}$$

for $n = 1, 2, 3, \ldots$ so that

$$u = \sum_{n=1}^{\infty} A_n e^{-\left(h+\frac{kn^2\pi^2}{L^2}\right)t} \sin\frac{n\pi}{L}x.$$

Imposing

$$u(x,0) = f(x) = \sum_{n=1}^{\infty} A_n \sin\frac{n\pi}{L}x$$

gives

$$u = \frac{2}{L}\sum_{n=1}^{\infty}\left(\int_0^L f(x)\sin\frac{n\pi}{L}x\,dx\right) e^{-\left(h+\frac{kn^2\pi^2}{L^2}\right)t}\sin\frac{n\pi}{L}x.$$

Exercises 12.4

3. Using $u = XT$ and $-\lambda^2$ as a separation constant leads to

$$X'' + \lambda^2 X = 0,$$
$$X(0) = 0,$$
$$X(L) = 0,$$

and

$$T'' + \lambda^2 a^2 T = 0.$$

Then

$$X = c_1 \sin\frac{n\pi}{L}x \quad \text{and} \quad T = c_2\cos\frac{n\pi a}{L}t + c_3 \sin\frac{n\pi a}{L}t$$

for $n = 1, 2, 3, \ldots$ so that

$$u = \sum_{n=1}^{\infty}\left(A_n \cos\frac{n\pi a}{L}t + B_n \sin\frac{n\pi a}{L}t\right)\sin\frac{n\pi}{L}x.$$

Imposing

$$u(x,0) = \sum_{n=1}^{\infty} A_n \sin\frac{n\pi}{L}x$$

gives

$$A_n = \frac{2}{L}\left(\int_0^{L/3}\frac{3}{L}x\sin\frac{n\pi}{L}x\,dx + \int_{L/3}^{2L/3}\sin\frac{n\pi}{L}x\,dx + \int_{2L/3}^{L}\left(3-\frac{3}{L}x\right)\sin\frac{n\pi}{L}x\,dx\right)$$

so that

$$A_1 = \frac{6\sqrt{3}}{\pi^2},$$
$$A_2 = A_3 = A_4 = 0,$$
$$A_5 = -\frac{6\sqrt{3}}{5^2\pi^2},$$
$$A_6 = 0,$$
$$a_7 = \frac{6\sqrt{3}}{7^2\pi^2}\ \ldots.$$

Imposing

$$u_t(x,0) = 0 = \sum_{n=1}^{\infty} B_n\frac{n\pi a}{L}\sin\frac{n\pi}{L}x$$

gives $B_n = 0$ for $n = 1, 2, 3, \ldots$ so that

$$u(x,t) = \frac{6\sqrt{3}}{\pi^2}\left(\cos\frac{\pi a}{L}t\,\sin\frac{\pi}{L}x - \frac{1}{5^2}\cos\frac{5\pi a}{L}t\,\sin\frac{5\pi}{L}x + \frac{1}{7^2}\cos\frac{7\pi a}{L}t\,\sin\frac{7\pi}{L}x - \cdots\right).$$

6. Using $u = XT$ and $-\lambda^2$ as a separation constant leads to

$$X'' + \lambda^2 X = 0,$$
$$X(0) = 0,$$
$$X(1) = 0,$$

and

$$T'' + \lambda^2 a^2 T = 0.$$

Then

$$X = c_1 \sin n\pi x \quad \text{and} \quad T = c_2 \cos n\pi at + c_3 \sin n\pi at$$

for $n = 1, 2, 3, \ldots$ so that

$$u = \sum_{n=1}^{\infty} (A_n \cos n\pi at + B_n \sin n\pi at)\sin n\pi x.$$

Imposing

$$u(x,0) = 0.01 \sin 3\pi x = \sum_{n=1}^{\infty} A_n \sin n\pi x$$

and

$$u_t(x,0) = 0 = \sum_{n=1}^{\infty} B_n n\pi a \, \sin n\pi x$$

gives $B_n = 0$ for $n = 1, 2, 3, \ldots$, $A_3 = 0.01$, and $A_n = 0$ for $n = 1, 2, 4, 5, 6, \ldots$ so that

$$u(x,t) = 0.01 \sin 3\pi x \, \cos 3\pi at.$$

9. Using $u = XT$ and $-\lambda^2$ as a separation constant leads to

$$X'' + \lambda^2 X = 0,$$
$$X(0) = 0,$$
$$X(\pi) = 0,$$

and

$$T'' + 2\beta T' + \lambda^2 T = 0.$$

Then

$$X = c_1 \sin nx \quad \text{and} \quad T = e^{-\beta t}\left(c_2 \cos\sqrt{n^2 - \beta^2}\,t + c_3 \sin\sqrt{n^2 - \beta^2}\,t\right)$$

so that

$$u = \sum_{n=1}^{\infty} e^{-\beta t}\left(A_n \cos\sqrt{n^2-\beta^2}\,t + B_n \sin\sqrt{n^2-\beta^2}\,t\right)\sin nx.$$

Imposing

$$u(x,0) = f(x) = \sum_{n=1}^{\infty} A_n \sin nx$$

and

$$u_t(x,0) = 0 = \sum_{n=1}^{\infty}\left(B_n\sqrt{n^2-\beta^2} - \beta A_n\right)\sin nx$$

gives

$$u(x,t) = e^{-\beta t}\sum_{n=1}^{\infty} A_n\left(\cos\sqrt{n^2-\beta^2}\,t + \frac{\beta}{\sqrt{n^2-\beta^2}}\sin\sqrt{n^2-\beta^2}\,t\right)\sin nx,$$

where

$$A_n = \frac{2}{\pi}\int_0^{\pi} f(x)\sin nx\,dx.$$

12. In this case the boundary conditions become, for $t > 0$,

$$u(0,t) = 0, \qquad u(L,t) = 0,$$

$$\left.\frac{\partial u}{\partial x}\right|_{x=0} = 0, \qquad \left.\frac{\partial u}{\partial x}\right|_{x=L} = 0.$$

15. $u(x,t) = \dfrac{1}{2}[\sin(x+at) + \sin(x-at)] + \dfrac{1}{2a}\displaystyle\int_{x-at}^{x+at} ds$

$$= \frac{1}{2}[\sin x\cos at + \cos x\sin at + \sin x\cos at - \cos x\sin at] + \frac{1}{2a}\,s\,\Big|_{x-at}^{x+at} = \sin x\cos at + t$$

18.

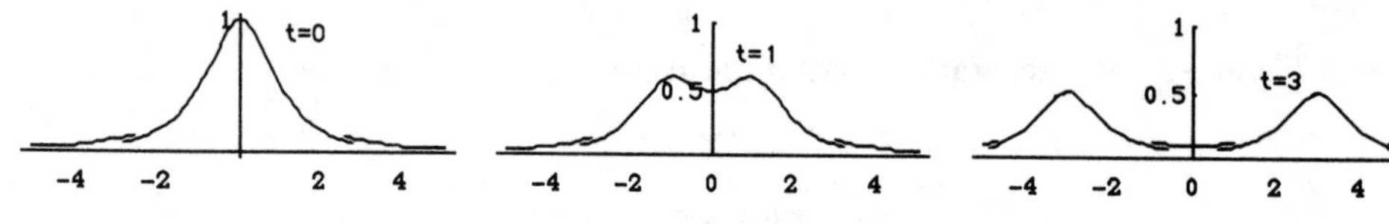

Exercises 12.5

3. Using $u = XY$ and $-\lambda^2$ as a separation constant leads to

$$X'' + \lambda^2 X = 0,$$

$$X(0) = 0,$$

$$X(a) = 0,$$

and

$$Y'' - \lambda^2 Y = 0,$$

$$Y(b) = 0.$$

Then

$$X = c_1 \sin\frac{n\pi}{a}x \quad \text{and} \quad Y = c_2 \cosh\frac{n\pi}{a}y - c_2\frac{\cosh\frac{n\pi b}{a}}{\sinh\frac{n\pi b}{a}}\sinh\frac{n\pi}{a}y$$

for $n = 1, 2, 3, \ldots$ so that

$$u = \sum_{n=1}^{\infty} A_n \sin\frac{n\pi}{a}x\left(\cosh\frac{n\pi}{a}y - \frac{\cosh\frac{n\pi b}{a}}{\sinh\frac{n\pi b}{a}}\sinh\frac{n\pi}{a}y\right).$$

Imposing

$$u(x,0) = f(x) = \sum_{n=1}^{\infty} A_n \sin\frac{n\pi}{a}x$$

gives

$$A_n = \frac{2}{a}\int_0^a f(x)\sin\frac{n\pi}{a}x\,dx$$

so that

$$u(x,y) = \frac{2}{a}\sum_{n=1}^{\infty}\left(\int_0^a f(x)\sin\frac{n\pi}{a}x\,dx\right)\sin\frac{n\pi}{a}x\left(\cosh\frac{n\pi}{a}y - \frac{\cosh\frac{n\pi b}{a}}{\sinh\frac{n\pi b}{a}}\sinh\frac{n\pi}{a}y\right).$$

6. Using $u = XY$ and λ^2 as a separation constant leads to

$$X'' - \lambda^2 X = 0,$$

$$X'(1) = 0,$$

and

$$Y'' + \lambda^2 Y = 0,$$

$$Y'(0) = 0,$$

$$Y'(\pi) = 0.$$

Then

$$Y = c_1 \cos ny$$

for $n = 0, 1, 2, \ldots$ and

$$X = c_2 \cosh nx - c_2 \frac{\sinh n}{\cosh n} \sinh nx$$

for $n = 0, 1, 2, \ldots$ so that

$$u = A_0 + \sum_{n=1}^{\infty} A_n \left(\cosh nx - \frac{\sinh n}{\cosh n} \sinh nx \right) \cos ny.$$

Imposing

$$u(0, y) = g(y) = A_0 + \sum_{n=1}^{\infty} A_n \cos ny$$

gives

$$A_0 = \frac{1}{\pi} \int_0^{\pi} g(y)\, dy \quad \text{and} \quad A_n = \frac{2}{\pi} \int_0^{\pi} g(y) \cos ny\, dy$$

for $n = 1, 2, 3, \ldots$ so that

$$u(x, y) = \frac{1}{\pi} \int_0^{\pi} g(y)\, dy + \sum_{n=1}^{\infty} \left(\frac{2}{\pi} \int_0^{\pi} g(y) \cos ny\, dy \right) \left(\cosh nx - \frac{\sinh n}{\cosh n} \sinh nx \right) \cos ny.$$

9. Using $u = XY$ and $-\lambda^2$ as a separation constant leads to

$$X'' + \lambda^2 X = 0,$$
$$X(0) = 0,$$
$$X(\pi) = 0,$$

and

$$Y'' - \lambda^2 Y = 0.$$

Then the boundedness of u as $y \to \infty$ gives $Y = c_1 e^{-ny}$ and $X = c_2 \sin nx$ for $n = 1, 2, 3, \ldots$ so that

$$u = \sum_{n=1}^{\infty} A_n e^{-ny} \sin nx.$$

Imposing

$$u(x, 0) = f(x) = \sum_{n=1}^{\infty} A_n \sin nx$$

gives

$$A_n = \frac{2}{\pi} \int_0^{\pi} f(x) \sin nx\, dx$$

so that

$$u(x, y) = \sum_{n=1}^{\infty} \left(\frac{2}{\pi} \int_0^{\pi} f(x) \sin nx\, dx \right) e^{-ny} \sin nx.$$

12. Since the boundary conditions at $x = 0$ and $x = a$ are functions of y we choose to separate Laplace's equation as

$$\frac{X''}{X} = -\frac{Y''}{Y} = \lambda^2$$

so that

$$X'' - \lambda^2 X = 0$$

$$Y'' + \lambda^2 Y = 0$$

and

$$X(x) = c_1 \cosh \lambda x + c_2 \sinh \lambda x$$

$$Y(y) = c_3 \cos \lambda y + c_2 \sin \lambda y.$$

Now $Y(0) = 0$ gives $c_3 = 0$ and $Y(b) = 0$ implies $\sin \lambda b = 0$ or $\lambda = n\pi/b$ for $n = 1, 2, 3, \ldots$. Thus

$$u_n(x, y) = XY = \left(A_n \cosh \frac{n\pi}{b} x + B_n \sinh \frac{n\pi}{b} x\right) \sin \frac{n\pi}{b} y$$

and

$$u(x, y) = \sum_{n=1}^{\infty} \left(A_n \cosh \frac{n\pi}{b} x + B_n \sinh \frac{n\pi}{b} x\right) \sin \frac{n\pi}{b} y. \tag{1}$$

At $x = 0$ we then have

$$F(y) = \sum_{n=1}^{\infty} A_n \sin \frac{n\pi}{b} y$$

and consequently

$$A_n = \frac{2}{b} \int_0^b F(y) \sin \frac{n\pi}{b} y \, dy. \tag{2}$$

At $x = a$,

$$G(y) = \sum_{n=1}^{\infty} \left(A_n \cosh \frac{n\pi}{b} a + B_n \sinh \frac{n\pi}{b} a\right) \sin \frac{n\pi}{b} y$$

indicates that the entire expression in the parentheses is given by

$$A_n \cosh \frac{n\pi}{b} a + B_n \sinh \frac{n\pi}{b} a = \frac{2}{b} \int_0^b G(y) \sin \frac{n\pi}{b} y \, dy.$$

We can now solve for B_n:

$$B_n \sinh \frac{n\pi}{b} a = \frac{2}{b} \int_0^b G(y) \sin \frac{n\pi}{b} y \, dy - A_n \cosh \frac{n\pi}{b} a$$

$$B_n = \frac{1}{\sinh \frac{n\pi}{b} a} \left(\frac{2}{b} \int_0^b G(y) \sin \frac{n\pi}{b} y \, dy - A_n \cosh \frac{n\pi}{b} a\right). \tag{3}$$

A solution to the given boundary-value problem consists of the series (1) with coefficients A_n and B_n given in (2) and (3), respectively.

Exercises 12.6

3. If we let $u(x,t) = v(x,t) + \psi(x)$, then we obtain as in Example 1 in the text

$$k\psi'' + r = 0$$

or

$$\psi(x) = -\frac{r}{2k}x^2 + c_1x + c_2.$$

The boundary conditions become

$$u(0,t) = v(0,t) + \psi(0) = u_0$$

$$u(1,t) = v(1,t) + \psi(1) = u_0.$$

Letting $\psi(0) = \psi(1) = u_0$ we obtain homogeneous boundary conditions in v:

$$v(0,t) = 0 \qquad \text{and} \qquad v(1,t) = 0.$$

Now $\psi(0) = \psi(1) = u_0$ implies $c_2 = u_0$ and $c_1 = r/2k$. Thus

$$\psi(x) = -\frac{r}{2k}x^2 + \frac{r}{2k}x + u_0 = u_0 - \frac{r}{2k}x(x-1).$$

To determine $v(x,t)$ we solve

$$k\frac{\partial^2 v}{\partial x^2} = \frac{\partial v}{dt}, \quad 0 < x < 1,\ t > 0$$

$$v(0,t) = 0, \quad v(1,t) = 0,$$

$$v(x,0) = \frac{r}{2k}x(x-1) - u_0.$$

Separating variables, we find

$$v(x,t) = \sum_{n=1}^{\infty} A_n e^{-kn^2\pi^2 t} \sin n\pi x,$$

where

$$A_n = 2\int_0^1 \left[\frac{r}{2k}x(x-1) - u_0\right] \sin n\pi x\, dx = 2\left[\frac{u_0}{n\pi} + \frac{r}{kn^3\pi^3}\right][(-1)^n - 1]. \tag{4}$$

Hence, a solution of the original problem is

$$u(x,t) = \psi(x) + v(x,t)$$

$$= u_0 - \frac{r}{2k}x(x-1) + \sum_{n=1}^{\infty} A_n e^{-kn^2\pi^2 t} \sin n\pi x,$$

where A_n is defined in (4).

6. Substituting $u(x,t) = v(x,t) + \psi(x)$ into the partial differential equation gives

$$k\frac{\partial^2 v}{\partial x^2} + k\psi'' - hv - h\psi = \frac{\partial v}{\partial t}.$$

This equation will be homogeneous provided ψ satisfies

$$k\psi'' - h\psi = 0.$$

Since k and h are positive, the general solution of this latter equation is

$$\psi(x) = c_1 \cosh\sqrt{\frac{h}{k}}\,x + c_2 \sinh\sqrt{\frac{h}{k}}\,x.$$

From $\psi(0) = 0$ and $\psi(\pi) = u_0$ we find $c_1 = 0$ and $c_2 = u_0/\sinh\sqrt{h/k}\,\pi$. Hence

$$\psi(x) = u_0\frac{\sinh\sqrt{h/k}\,x}{\sinh\sqrt{h/k}\,\pi}.$$

Now the new problem is

$$k\frac{\partial^2 v}{\partial x^2} - hv = \frac{\partial v}{\partial t}, \quad 0 < x < \pi,\ t > 0$$

$$v(0,t) = 0, \quad v(\pi,t) = 0, \quad t > 0$$

$$v(x,0) = -\psi(x), \quad 0 < x < \pi.$$

If we let $v = XT$ then

$$\frac{X''}{X} = \frac{T' + hT}{kT} = -\lambda^2$$

gives the separated differential equations

$$X'' + \lambda^2 X = 0 \qquad \text{and} \qquad T' + \left(h + k\lambda^2\right)T = 0.$$

The respective solutions are

$$X(x) = c_3\cos\lambda x + c_4\sin\lambda x$$

$$T(t) = c_5 e^{-(h+k\lambda^2)t}.$$

From $X(0) = 0$ we get $c_3 = 0$ and from $X(\pi) = 0$ we find $\lambda = n$ for $n = 1, 2, 3, \ldots$. Consequently, it follows that

$$v(x,t) = \sum_{n=1}^{\infty} A_n e^{-(h+kn^2)t}\sin nx$$

where

$$A_n = -\frac{2}{\pi}\int_0^{\pi}\psi(x)\sin nx\,dx.$$

Hence a solution of the original problem is

$$u(x,t) = u_0 \frac{\sinh\sqrt{h/k}\,x}{\sinh\sqrt{h/k}\,\pi} + e^{-ht}\sum_{n=1}^{\infty} A_n e^{-kn^2t}\sin nx$$

where

$$A_n = -\frac{2}{\pi}\int_0^{\pi} u_0 \frac{\sinh\sqrt{h/k}\,x}{\sinh\sqrt{h/k}\,\pi}\sin nx\,dx.$$

Using the exponential definition of the hyperbolic sine and integration by parts we find

$$A_n = \frac{2u_0nk(-1)^n}{\pi\,(h+kn^2)}.$$

9. Substituting $u(x,t) = v(x,t) + \psi(x)$ into the partial differential equation gives

$$a^2\frac{\partial^2 v}{\partial x^2} + a^2\psi'' + Ax = \frac{\partial^2 v}{\partial t^2}.$$

This equation will be homogeneous provided ψ satisfies

$$a^2\psi'' + Ax = 0.$$

The general solution of this differential equation is

$$\psi(x) = -\frac{A}{6a^2}x^3 + c_1x + c_2.$$

From $\psi(0) = 0$ we obtain $c_2 = 0$, and from $\psi(1) = 0$ we obtain $c_1 = A/6a^2$. Hence

$$\psi(x) = \frac{A}{6a^2}(x - x^3).$$

Now the new problem is

$$a^2\frac{\partial^2 v}{\partial x^2} = \frac{\partial^2 v}{\partial t^2}$$

$$v(0,t) = 0, \quad v(1,t) = 0, \quad t > 0,$$

$$v(x,0) = -\psi(x), \quad v_t(x,0) = 0, \quad 0 < x < 1.$$

Identifying this as the wave equation solved in Section 12.4 in the text with $L = 1$, $f(x) = -\psi(x)$, and $g(x) = 0$ we obtain

$$v(x,t) = \sum_{n=1}^{\infty} A_n\cos n\pi at\sin n\pi x$$

where

$$A_n = 2\int_0^1[-\psi(x)]\sin n\pi x\,dx = \frac{A}{3a^2}\int_0^1(x^3 - x)\sin n\pi x\,dx = \frac{2A(-1)^n}{a^2\pi^3n^3}.$$

Thus

$$u(x,t) = \frac{A}{6a^2}(x - x^3) + \frac{2A}{a^2\pi^3} \sum_{n=1}^{\infty} \frac{(-1)^n}{n^3} \cos n\pi at \sin n\pi x.$$

12. Substituting $u(x,y) = v(x,y) + \psi(x)$ into Poisson's equation we obtain

$$\frac{\partial^2 v}{\partial x^2} + \psi''(x) + h + \frac{\partial^2 v}{\partial y^2} = 0.$$

The equation will be homogeneous provided ψ satisfies $\psi''(x) + h = 0$ or $\psi(x) = -\frac{h}{2}x^2 + c_1 x + c_2$. From $\psi(0) = 0$ we obtain $c_2 = 0$. From $\psi(\pi) = 1$ we obtain

$$c_1 = \frac{1}{\pi} + \frac{h\pi}{2}.$$

Then

$$\psi(x) = \left(\frac{1}{\pi} + \frac{h\pi}{2}\right)x - \frac{h}{2}x^2.$$

The new boundary-value problem is

$$\frac{\partial^2 v}{\partial x^2} + \frac{\partial^2 v}{\partial y^2} = 0$$

$$v(0,y) = 0, \quad v(\pi,y) = 0,$$

$$v(x,0) = -\psi(x), \quad 0 < x < \pi.$$

This is Problem 9 in Section 12.5. The solution is

$$v(x,y) = \sum_{n=1}^{\infty} A_n e^{-ny} \sin nx$$

where

$$A_n = \frac{2}{\pi}\int_0^{\pi} [-\psi(x) \sin nx]\, dx$$

$$= \frac{2(-1)^n}{m}\left(\frac{1}{\pi} + \frac{h\pi}{2}\right) - h(-1)^n\left(\frac{\pi}{n} + \frac{2}{n^2}\right).$$

Thus

$$u(x,y) = v(x,y) + \psi(x) = \left(\frac{1}{\pi} + \frac{h\pi}{2}\right)x - \frac{h}{2}x^2 + \sum_{n=1}^{\infty} A_n e^{-ny} \sin nx.$$

Exercises 12.7

3. Separating variables in Laplace's equation gives

$$X'' + \lambda^2 X = 0$$

$$Y'' - \lambda^2 Y = 0$$

and

$$X(x) = c_1 \cos \lambda x + c_2 \sin \lambda x$$

$$Y(y) = c_3 \cosh \lambda y + c_4 \sinh \lambda y.$$

From $u(0, y) = 0$ we obtain $X(0) = 0$ and $c_1 = 0$. From $u_x(a, y) = -hu(a, y)$ we obtain $X'(a) = -hX(a)$ and

$$\lambda \cos \lambda a = -h \sin \lambda a \quad \text{or} \quad \tan \lambda a = -\frac{\lambda}{h}.$$

Let λ_n, where $n = 1, 2, 3, \ldots$, be the consecutive positive roots of this equation. From $u(x, 0) = 0$ we obtain $Y(0) = 0$ and $c_3 = 0$. Thus

$$u(x, y) = \sum_{n=1}^{\infty} A_n \sinh \lambda_n y \sin \lambda_n x.$$

Now

$$f(x) = \sum_{n=1}^{\infty} A_n \sinh \lambda_n b \sin \lambda_n x$$

and

$$A_n \sinh \lambda_n b = \frac{\int_0^a f(x) \sin \lambda_n x \, dx}{\int_0^a \sin^2 \lambda_n x \, dx}.$$

Since

$$\begin{aligned}
\int_0^a \sin^2 \lambda_n x \, dx &= \frac{1}{2}\left[a - \frac{1}{2\lambda_n} \sin 2\lambda_n a\right] = \frac{1}{2}\left[a - \frac{1}{\lambda_n} \sin \lambda_n a \cos \lambda_n a\right] \\
&= \frac{1}{2}\left[a - \frac{1}{h\lambda_n}(h \sin \lambda_n a) \cos \lambda_n a\right] \\
&= \frac{1}{2}\left[a - \frac{1}{h\lambda_n}(-\lambda_n \cos \lambda_n a) \cos \lambda_n a\right] = \frac{1}{2h}\left[ah + \cos^2 \lambda_n a\right],
\end{aligned}$$

we have

$$A_n = \frac{2h}{\sinh \lambda_n b[ah + \cos^2 \lambda_n a]} \int_0^a f(x) \sin \lambda_n x \, dx.$$

6. Substituting $u(x, t) = v(x, t) + \psi(x)$ into the partial differential equation gives

$$a^2 \frac{\partial^2 v}{\partial x^2} + \psi''(x) = \frac{\partial^2 v}{\partial t^2}.$$

This equation will be homogeneous if $\psi''(x) = 0$ or $\psi(x) = c_1x + c_2$. The boundary condition $u(0,t) = 0$ implies $\psi(0) = 0$ which implies $c_2 = 0$. Thus $\psi(x) = c_1x$. Using the second boundary condition, we obtain

$$E\left(\frac{\partial v}{\partial x} + \psi'\right)\Big|_{x=L} = F_0,$$

which will be homogeneous when

$$E\psi'(L) = F_0.$$

Since $\psi'(x) = c_1$ we conclude that $c_1 = F_0/E$ and

$$\psi(x) = \frac{F_0}{E}x.$$

The new boundary-value problem is

$$a^2\frac{\partial^2 v}{\partial x^2} = \frac{\partial^2 v}{\partial t^2}, \quad 0 < x < L, \quad t > 0$$

$$v(0,t) = 0, \quad \frac{\partial v}{\partial x}\Big|_{x=L} = 0, \quad t > 0,$$

$$v(x,0) = -\frac{F_0}{E}x, \quad \frac{\partial v}{\partial t}\Big|_{t=0} = 0, \quad 0 < x < L.$$

Referring to Example 2 in the text we see that

$$v(x,t) = \sum_{n=1}^{\infty} A_n \cos a\left(\frac{2n-1}{2L}\right)\pi t \sin\left(\frac{2n-1}{2L}\right)\pi x$$

where

$$-\frac{F_0}{E}x = \sum_{n=1}^{\infty} A_n \sin\left(\frac{2n-1}{2L}\right)\pi x$$

and

$$A_n = \frac{-F_0\int_0^L x\sin\left(\frac{2n-1}{2L}\right)\pi x\,dx}{E\int_0^L \sin^2\left(\frac{2n-1}{2L}\right)\pi x\,dx} = \frac{8F_0L(-1)^n}{E\pi^2(2n-1)^2}.$$

Thus

$$\begin{aligned} u(x,t) &= v(x,t) + \psi(x) \\ &= \frac{F_0}{E}x + \frac{8F_0L}{E\pi^2}\sum_{n=1}^{\infty}\frac{(-1)^n}{(2n-1)^2}\cos a\left(\frac{2n-1}{2L}\right)\pi t \sin\left(\frac{2n-1}{2L}\right)\pi x. \end{aligned}$$

9. (a) Using $u = XT$ and separation constant λ^4 we find

$$X^{(4)} - \lambda^4 X = 0$$

and

$$X(x) = c_1\cos\lambda x + c_2\sin\lambda x + c_3\cosh\lambda x + c_4\sinh\lambda x.$$

Since $u = XT$ the boundary conditions become

$$X(0) = 0, \quad X'(0) = 0, \quad X''(1) = 0, \quad X'''(1) = 0.$$

Now $X(0) = 0$ implies $c_1 + c_3 = 0$, while $X'(0) = 0$ implies $c_2 + c_4 = 0$. Thus

$$X(x) = c_1 \cos \lambda x + c_2 \sin \lambda x - c_1 \cosh \lambda x - c_2 \sinh \lambda x.$$

The boundary condition $X''(1) = 0$ implies

$$-c_1 \cos \lambda - c_2 \sin \lambda - c_1 \cosh \lambda - c_2 \sinh \lambda = 0$$

while the boundary condition $X'''(1) = 0$ implies

$$c_1 \sin \lambda - c_2 \cos \lambda - c_1 \sinh \lambda - c_2 \cosh \lambda = 0.$$

We then have the system of two equations in two unknowns

$$(\cos \lambda + \cosh \lambda)c_1 + (\sin \lambda + \sinh \lambda)c_2 = 0$$

$$(\sin \lambda - \sinh \lambda)c_1 - (\cos \lambda + \cosh \lambda)c_2 = 0.$$

This homogeneous system will have nontrivial solutions for c_1 and c_2 provided

$$\begin{vmatrix} \cos \lambda + \cosh \lambda & \sin \lambda + \sinh \lambda \\ \sin \lambda - \sinh \lambda & -\cos \lambda - \cosh \lambda \end{vmatrix} = 0$$

or

$$-2 - 2\cos \lambda \cosh \lambda = 0.$$

Thus, the eigenvalues are determined by the equation $\cos \lambda \cosh \lambda = -1$.

(b) Using a computer to graph $\cosh \lambda$ and $-1/\cos \lambda = -\sec \lambda$ we see that the first two positive eigenvalues occur near 1.9 and 4.7. Applying Newton's method with these initial values we find that the eigenvalues are $\lambda_1 = 1.8751$ and $\lambda_2 = 4.6941$.

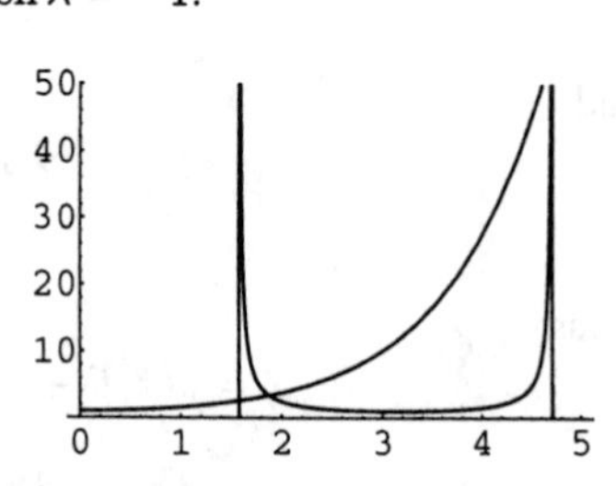

Exercises 12.8

3. In this problem we need to solve the partial differential equation

$$a^2\left(\frac{\partial^2 u}{\partial x^2}+\frac{\partial^2 u}{\partial y^2}\right)=\frac{\partial^2 u}{\partial t^2}.$$

To separate this equation we try $u(x,y,t)=X(x)Y(y)T(t)$:

$$a^2(X''YT+XY''T)=XYT''$$

$$\frac{X''}{X}=-\frac{Y''}{Y}+\frac{T''}{a^2T}=-\lambda^2.$$

Then

$$X''+\lambda^2X=0 \tag{5}$$

$$\frac{Y''}{Y}=\frac{T''}{a^2T}+\lambda^2=-\mu^2$$

$$Y''+\mu^2Y=0 \tag{6}$$

$$T''+a^2\left(\lambda^2+\mu^2\right)T=0. \tag{7}$$

The general solutions of equations (5), (6), and (7) are, respectively,

$$X(x)=c_1\cos\lambda x+c_2\sin\lambda x$$

$$Y(y)=c_3\cos\mu y+c_4\sin\mu y$$

$$T(t)=c_5\cos a\sqrt{\lambda^2+\mu^2}\,t+c_6\sin a\sqrt{\lambda^2+\mu^2}\,t.$$

The conditions $X(0)=0$ and $Y(0)=0$ give $c_1=0$ and $c_3=0$. The conditions $X(\pi)=0$ and $Y(\pi)=0$ yield two sets of eigenvalues:

$$\lambda=m,\ m=1,2,3,\dots \quad\text{and}\quad \mu=n,\ n=1,2,3,\dots.$$

A product solution of the partial differential equation that satisfies the boundary conditions is

$$u_{mn}(x,y,t)=\left(A_{mn}\cos a\sqrt{m^2+n^2}\,t+B_{mn}\sin a\sqrt{m^2+n^2}\,t\right)\sin mx\sin ny.$$

To satisfy the initial conditions we use the superposition principle:

$$u(x,y,t)=\sum_{m=1}^{\infty}\sum_{n=1}^{\infty}\left(A_{mn}\cos a\sqrt{m^2+n^2}\,t+B_{mn}\sin a\sqrt{m^2+n^2}\,t\right)\sin mx\sin ny.$$

The initial condition $u_t(x,y,0)=0$ implies $B_{mn}=0$ and

$$u(x,y,t)=\sum_{m=1}^{\infty}\sum_{n=1}^{\infty}A_{mn}\cos a\sqrt{m^2+n^2}\,t\sin mx\sin ny.$$

At $t = 0$ we have

$$xy(x-\pi)(y-\pi) = \sum_{m=1}^{\infty}\sum_{n=1}^{\infty} A_{mn} \sin mx \sin ny.$$

It follows that (See Problem 52, Exercises 11.3)

$$\begin{aligned} A_{mn} &= \frac{4}{\pi^2}\int_0^{\pi}\int_0^{\pi} xy(x-\pi)(y-\pi)\sin mx \sin ny\, dx\, dy \\ &= \frac{4}{\pi^2}\int_0^{\pi} x(x-\pi)\sin mx\, dx \int_0^{\pi} y(y-\pi)\sin ny\, dy \\ &= \frac{16}{m^3 n^3 \pi^2}[(-1)^m - 1][(-1)^n - 1]. \end{aligned}$$

6. To separate Laplace's equation in three dimensions we try $u(x,y,z) = X(x)Y(y)Z(z)$:

$$X''YZ + XY''Z + XYZ'' = 0$$

$$\frac{X''}{X} = -\frac{Y''}{Y} - \frac{Z''}{Z} = -\lambda^2.$$

Then

$$X'' + \lambda^2 X = 0 \tag{8}$$

$$\frac{Y''}{Y} = -\frac{Z''}{Z} + \lambda^2 = -\mu^2$$

$$Y'' + \mu^2 Y = 0 \tag{9}$$

$$Z'' - (\lambda^2 + \mu^2)Z = 0. \tag{10}$$

The general solutions of equations (8), (9), and (10) are, respectively

$$X(x) = c_1 \cos \lambda x + c_2 \sin \lambda x$$

$$Y(y) = c_3 \cos \mu y + c_4 \sin \mu y$$

$$Z(z) = c_5 \cosh \sqrt{\lambda^2 + \mu^2}\, z + c_6 \sinh \sqrt{\lambda^2 + \mu^2}\, z.$$

The boundary and initial conditions are

$$u(0,y,z) = 0, \qquad u(a,y,z) = 0,$$

$$u(x,0,z) = 0, \qquad u(x,b,z) = 0,$$

$$u(x,y,0) = f(x,y), \qquad u(x,y,c) = 0.$$

The conditions $X(0) = Y(0) = 0$ give $c_1 = c_3 = 0$. The conditions $X(a) = Y(b) = 0$ yield two sets of eigenvalues:

$$\lambda = \frac{m\pi}{a}, \; m = 1,2,3,\ldots \quad \text{and} \quad \mu = \frac{n\pi}{b}, \; n = 1,2,3,\ldots.$$

Let

$$\omega_{mn}^2 = \frac{m^2\pi^2}{a^2} + \frac{n^2\pi^2}{b^2}.$$

Then the boundary condition $Z(c) = 0$ gives

$$c_5 \cosh c\omega_{mn} + c_6 \sinh c\omega_{mn} = 0$$

from which we obtain

$$\begin{aligned} Z(z) &= c_5\left(\cosh w_{mn}z - \frac{\cosh c\omega_{mn}}{\sinh c\omega_{mn}}\sinh \omega z\right) \\ &= \frac{c_5}{\sinh c\omega_{mn}}(\sinh c\omega_{mn}\cosh \omega_{mn}z - \cosh c\omega_{mn}\sinh \omega_{mn}z) \\ &= c_{mn}\sinh \omega_{mn}(c-z). \end{aligned}$$

By the superposition principle

$$u(x,y,t) = \sum_{m=1}^{\infty}\sum_{n=1}^{\infty} A_{mn}\sinh \omega_{mn}(c-z)\sin\frac{m\pi}{a}x\sin\frac{n\pi}{b}y$$

where

$$A_{mn} = \frac{4}{ab\sinh c\omega_{mn}}\int_0^b\int_0^a f(x,y)\sin\frac{m\pi}{a}x\sin\frac{n\pi}{b}y\,dx\,dy.$$

Chapter 12 Review Exercises

3. Substituting $u(x,t) = v(x,t) + \psi(x)$ into the partial differential equation we obtain

$$k\frac{\partial^2 v}{\partial x^2} + k\psi''(x) = \frac{\partial v}{\partial t}.$$

This equation will be homogeneous provided ψ satisfies

$$k\psi'' = 0 \quad \text{or} \quad \psi = c_1x + c_2.$$

Considering

$$u(0,t) = v(0,t) + \psi(0) = u_0$$

we set $\psi(0) = u_0$ so that $\psi(x) = c_1x + u_0$. Now

$$-\left.\frac{\partial u}{\partial x}\right|_{x=\pi} = -\left.\frac{\partial v}{\partial x}\right|_{x=\pi} - \psi'(x) = v(\pi,t) + \psi(\pi) - u_1$$

is equivalent to

$$\left.\frac{\partial v}{\partial x}\right|_{x=\pi} + v(\pi,t) = u_1 - \psi'(x) - \psi(\pi) = u_1 - c_1 - (c_1\pi + u_0),$$

which will be homogeneous when

$$u_1 - c_1 - c_1\pi - u_0 = 0 \quad \text{or} \quad c_1 = \frac{u_1 - u_0}{1+\pi}.$$

The steady-state solution is

$$\psi(x) = \left(\frac{u_1 - u_0}{1+\pi}\right)x + u_0.$$

6. The boundary-value problem is

$$\frac{\partial^2 u}{\partial x^2} + x^2 = \frac{\partial^2 u}{\partial t^2}, \quad 0 < x < 1, \quad t > 0,$$

$$u(0,t) = 1, \quad u(1,t) = 0, \quad t > 0,$$

$$u(x,0) = f(x), \quad u_t(x,0) = 0, \quad 0 < x < 1.$$

Substituting $u(x,t) = v(x,t) + \psi(x)$ into the partial differential equation gives

$$\frac{\partial^2 v}{\partial x^2} + \psi''(x) + x^2 = \frac{\partial^2 v}{\partial t^2}.$$

This equation will be homogeneous provided $\psi''(x) + x^2 = 0$ or

$$\psi(x) = -\frac{1}{12}x^4 + c_1 x + c_2.$$

From $\psi(0) = 1$ and $\psi(1) = 0$ we obtain $c_1 = -11/12$ and $c_2 = 1$. The new problem is

$$\frac{\partial^2 v}{\partial x^2} = \frac{\partial^2 v}{\partial t^2}, \quad 0 < x < 1, \quad t > 0,$$

$$v(0,t) = 0, \quad v(1,t) = 0, \quad t > 0,$$

$$v(x,0) = f(x) - \psi(x), \quad v_t(x,0) = 0, \quad 0 < x < 1.$$

From Section 12.4 in the text we see that $B_n = 0$,

$$A_n = 2\int_0^1 [f(x) - \psi(x)] \sin n\pi x \, dx = 2\int_0^1 \left[f(x) + \frac{1}{12}x^4 + \frac{11}{12}x - 1\right] \sin n\pi x \, dx,$$

and

$$v(x,t) = \sum_{n=1}^{\infty} A_n \cos n\pi t \sin n\pi x.$$

Thus

$$u(x,t) = v(x,t) + \psi(x) = -\frac{1}{12}x^4 - \frac{11}{12}x + 1 + \sum_{n=1}^{\infty} A_n \cos n\pi t \sin n\pi x.$$

9. Using $u = XY$ and λ^2 as a separation constant leads to

$$X'' - \lambda^2 X = 0,$$

and

$$Y'' + \lambda^2 Y = 0,$$
$$Y(0) = 0,$$
$$Y(\pi) = 0.$$

Then

$$Y = c_1 \sin ny \quad \text{and} \quad X = c_2 e^{-nx}$$

for $n = 1, 2, 3, \ldots$(since u must be bounded as $x \to \infty$) so that

$$u = \sum_{n=1}^{\infty} A_n e^{-nx} \sin ny.$$

Imposing

$$u(0, y) = 50 = \sum_{n=1}^{\infty} A_n \sin ny$$

gives

$$A_n = \frac{2}{\pi} \int_0^{\pi} 50 \sin ny \, dy = \frac{100}{n\pi}[1 - (-1)^n]$$

so that

$$u(x, y) = \sum_{n=1}^{\infty} \frac{100}{n\pi}[1 - (-1)^n]e^{-nx} \sin ny.$$

12. Substituting $u(x,t) = v(x,t) + \psi(x)$ into the partial differential equation gives

$$k\frac{\partial^2 v}{\partial x^2} + k\psi'' + \sin 2\pi x = \frac{\partial v}{\partial t}.$$

This equation will be homogeneous provided ψ satisfies

$$k\psi'' + \sin 2\pi x = 0.$$

The general solution of this equation is

$$\psi(x) = \frac{1}{4k\pi^2} \sin 2\pi x + c_1 x + c_2.$$

From $\psi(0) = \psi(1) = 0$ we find that $c_1 = c_2 = 0$ and

$$\psi(x) = \frac{1}{4k\pi^2} \sin 2\pi x.$$

Now the new problem is

$$k\frac{\partial^2 v}{\partial x^2} = \frac{\partial v}{\partial t}, \quad 0 < x < 1, \quad t > 0$$
$$v(0,t) = 0, \quad v(1,t) = 0, \quad t > 0$$
$$v(x,0) = \sin \pi x - \psi(x), \quad 0 < x < 1.$$

If we let $v = XT$ then

$$\frac{X''}{X} = \frac{T'}{kT} = -\lambda^2$$

gives the separated differential equations

$$X'' + \lambda^2 X = 0 \quad \text{and} \quad T' + k\lambda^2 T = 0.$$

The respective solutions are

$$X(x) = c_3 \cos \lambda x + c_4 \sin \lambda x$$

$$T(t) = c_5 e^{-k\lambda^2 t}.$$

From $X(0) = 0$ we get $c_3 = 0$ and from $X(1) = 0$ we find $\lambda = n\pi$ for $n = 1, 2, 3, \ldots$. Consequently, it follows that

$$v(x,t) = \sum_{n=1}^{\infty} A_n e^{-kn^2\pi^2 t} \sin n\pi x$$

where

$$v(x,0) = \sin \pi x - \frac{1}{4k\pi^2} \sin 2\pi x = 0$$

implies

$$A_n = 2\int_0^1 \left(\sin \pi x - \frac{1}{4k\pi^2} \sin 2\pi x\right) \sin n\pi x \, dx.$$

By orthogonality $A_n = 0$ for $n = 3, 4, 5, \ldots$, and only A_1 and A_2 can be nonzero. We have

$$A_1 = 2\left[\int_0^1 \sin^2 \pi x \, dx - \frac{1}{4k\pi^2}\int_0^1 \sin 2\pi x \sin \pi x \, dx\right] = 2\int_0^1 \frac{1}{2}(1 - \cos 2\pi x)\, dx = 1$$

and

$$A_2 = 2\left[\int_0^1 \sin \pi x \sin 2\pi x \, dx - \frac{1}{4k\pi^2}\int_0^1 \sin^2 2\pi x \, dx\right]$$

$$= -\frac{1}{2k\pi^2}\int_0^1 \frac{1}{2}(1 - \cos 4\pi x)\, dx = -\frac{1}{4k\pi^2}.$$

Therefore

$$v(x,t) = A_1 e^{-k\pi^2 t} \sin \pi x + A_2 e^{-k4\pi^2 t} \sin 2\pi x$$

$$= e^{-k\pi^2 t} \sin \pi x - \frac{1}{4k\pi^2} e^{-4k\pi^2 t} \sin 2\pi x$$

and

$$u(x,t) = v(x,t) + \psi(x) = e^{-k\pi^2 t} \sin \pi x + \frac{1}{4k\pi^2}(1 - e^{-4k\pi^2 t}) \sin 2\pi x.$$

13 Boundary-Value Problems in Other Coordinate Systems

Exercises 13.1

3. We have

$$A_0 = \frac{1}{2\pi}\int_0^{2\pi}(2\pi\theta - \theta^2)\,d\theta = \frac{2\pi^2}{3}$$

$$A_n = \frac{1}{\pi}\int_0^{2\pi}(2\pi\theta - \theta^2)\cos n\theta\,d\theta = -\frac{4}{n^2}$$

$$B_n = \frac{1}{\pi}\int_0^{2\pi}(2\pi\theta - \theta^2)\sin n\theta\,d\theta = 0$$

and so

$$u(r,\theta) = \frac{2\pi^2}{3} - 4\sum_{n=1}^{\infty}\frac{r^n}{n^2}\cos n\theta.$$

6. We solve

$$\frac{\partial^2 u}{\partial r^2} + \frac{1}{r}\frac{\partial u}{\partial r} + \frac{1}{r^2}\frac{\partial^2 u}{\partial \theta^2} = 0, \quad 0 < \theta < \frac{\pi}{2}, \quad 0 < r < c,$$

$$u(c,\theta) = f(\theta), \quad 0 < \theta < \frac{\pi}{2},$$

$$u(r,0) = 0, \quad u(r,\pi/2) = 0, \quad 0 < r < c.$$

Proceeding as in Example 1 of Section 13.1 in the text we obtain the separated differential equations

$$r^2R'' + rR' - \lambda^2 R = 0$$

$$\Theta'' + \lambda^2\Theta = 0$$

with solutions

$$\Theta(\theta) = c_1\cos\lambda\theta + c_2\sin\lambda\theta$$

$$R(r) = c_3r^{\lambda} + c_4r^{-\lambda}.$$

Since we want $R(r)$ to be bounded as $r \to 0$ we require $c_4 = 0$. Applying the boundary conditions $\Theta(0) = 0$ and $\Theta(\pi/2) = 0$ we find that $c_1 = 0$ and $\lambda = 2n$ for $n = 1, 2, 3, \ldots$. Therefore

$$u(r,\theta) = \sum_{n=1}^{\infty} A_n r^{2n}\sin 2n\theta.$$

From
$$u(c,\theta) = f(\theta) = \sum_{n=1}^{\infty} A_n c^n \sin 2n\theta$$

we find
$$A_n = \frac{4}{\pi c^{2n}} \int_0^{\pi/2} f(\theta) \sin 2n\theta \, d\theta.$$

9. Proceeding as in Example 1 of Section 13.1 and again using the periodicity of $u(r,\theta)$, we have
$$\Theta(\theta) = c_1 \cos \lambda\theta + c_2 \sin \lambda\theta$$

where $\lambda = n$ for $n = 0, 1, 2, \ldots$. Then
$$R(r) = c_3 r^n + c_4 r^{-n}.$$

[We do not have $c_4 = 0$ in this case since $0 < a \leq r$.] Since $u(b,\theta) = 0$ we have
$$u(r,\theta) = A_0 \ln \frac{r}{b} + \sum_{n=1}^{\infty} \left[\left(\frac{b}{r}\right)^n - \left(\frac{r}{b}\right)^n \right] [A_n \cos n\theta + B_n \sin n\theta].$$

From
$$u(a,\theta) = f(\theta) = A_0 \ln \frac{a}{b} + \sum_{n=1}^{\infty} \left[\left(\frac{b}{a}\right)^n - \left(\frac{a}{b}\right)^n \right] [A_n \cos n\theta + B_n \sin n\theta]$$

we find
$$A_0 \ln \frac{a}{b} = \frac{1}{2\pi} \int_0^{2\pi} f(\theta)\, d\theta,$$
$$\left[\left(\frac{b}{a}\right)^n - \left(\frac{a}{b}\right)^n \right] A_n = \frac{1}{\pi} \int_0^{2\pi} f(\theta) \cos n\theta \, d\theta,$$

and
$$\left[\left(\frac{b}{a}\right)^n - \left(\frac{a}{b}\right)^n \right] B_n = \frac{1}{\pi} \int_0^{2\pi} f(\theta) \sin n\theta \, d\theta.$$

12. Letting $u(r,\theta) = v(r,\theta) + \psi(\theta)$ we obtain $\psi''(\theta) = 0$ and so $\psi(\theta) = c_1\theta + c_2$. From $\psi(0) = 0$ and $\psi(\pi) = u_0$ we find, in turn, $c_2 = 0$ and $c_1 = u_0/\pi$. Therefore $\psi(\theta) = \dfrac{u_0}{\pi}\theta$. Now $u(1,\theta) = v(1,\theta) + \psi(\theta)$ so that $v(1,\theta) = u_0 - \dfrac{u_0}{\pi}\theta$.

From
$$v(r,\theta) = \sum_{n=1}^{\infty} A_n r^n \sin n\theta \quad \text{and} \quad v(1,\theta) = \sum_{n=1}^{\infty} A_n \sin n\theta$$

we obtain
$$A_n = \frac{2}{\pi} \int_0^{\pi} \left(u_0 - \frac{u_0}{\pi}\theta \right) \sin n\theta \, d\theta = \frac{2u_0}{\pi n}.$$

Thus
$$u(r,\theta) = \frac{u_0}{\pi}\theta + \frac{2u_0}{\pi} \sum_{n=1}^{\infty} \frac{r^n}{n} \sin n\theta.$$

Exercises 13.2

3. Referring to Example 2 in the text we have

$$R(r) = c_1 J_0(\lambda r) + c_2 Y_0(\lambda r)$$

$$Z(z) = c_3 \cosh \lambda z + c_4 \sinh \lambda z$$

where $c_2 = 0$ and $J_0(2\lambda) = 0$ defines the positive eigenvalues λ_n. From $Z(4) = 0$ we obtain

$$c_3 \cosh 4\lambda_n + c_4 \sinh 4\lambda_n = 0 \quad \text{or} \quad c_4 = -c_3 \frac{\cosh 4\lambda_n}{\sinh 4\lambda_n}.$$

Then

$$Z(z) = c_3 \left[\cosh \lambda_n z - \frac{\cosh 4\lambda_n}{\sinh 4\lambda_n} \sinh \lambda_n z\right] = c_3 \frac{\sinh 4\lambda_n \cosh \lambda_n z - \cosh 4\lambda_n \sinh \lambda_n z}{\sinh 4\lambda_n}$$

$$= c_3 \frac{\sinh \lambda_n (4 - z)}{\sinh 4\lambda_n}$$

and

$$u(r, z) = \sum_{n=1}^{\infty} A_n \frac{\sinh \lambda_n (4 - z)}{\sinh 4\lambda_n} J_0(\lambda_n r).$$

From

$$u(r, 0) = u_0 = \sum_{n=1}^{\infty} A_n J_0(\lambda_n r)$$

we obtain

$$A_n = \frac{2u_0}{4J_1^2(2\lambda_n)} \int_0^2 r J_0(\lambda_n r)\, dr = \frac{u_0}{\lambda_n J_1(2\lambda_n)}.$$

Thus the temperature in the cylinder is

$$u(r, z) = u_0 \sum_{n=1}^{\infty} \frac{\sinh \lambda_n (4 - z) J_0(\lambda_n r)}{\lambda_n \sinh 4\lambda_n J_1(2\lambda_n)}.$$

6. If the edge $r = c$ is insulated we have the boundary condition $u_r(c, t) = 0$. Letting $u(r, t) = R(r)T(t)$ and separating variables we obtain

$$\frac{R'' + \frac{1}{r} R'}{R} = \frac{T'}{kT} = \mu \quad \text{and} \quad R'' + \frac{1}{r} R' - \mu R = 0, \quad T' - \mu k T = 0.$$

From the second equation we find $T(t) = e^{\mu k t}$. If $\mu > 0$, $T(t)$ increases without bound as $t \to \infty$. Thus we assume $\mu = -\lambda^2 \leq 0$. Now

$$R'' + \frac{1}{r} R' + \lambda^2 R = 0$$

is a parametric Bessel equation with solution

$$R(r) = c_1 J_0(\lambda r) + c_2 Y_0(\lambda r).$$

Since Y_0 is unbounded as $r \to 0$ we take $c_2 = 0$. Then $R(r) = c_1 J_0(\lambda r)$ and the boundary condition $u_r(c,t) = R'(c)T(t) = 0$ implies

$$R'(c) = \lambda c_1 J_0'(\lambda c) = 0.$$

This defines an eigenvalue $\lambda = 0$ and positive eigenvalues λ_n. Thus

$$u(r,t) = A_0 + \sum_{n=1}^{\infty} A_n J_0(\lambda_n r) e^{-\lambda_n^2 kt}.$$

From

$$u(r,0) = f(r) = A_0 + \sum_{n=1}^{\infty} A_n J_0(\lambda_n r)$$

we find

$$A_0 = \frac{2}{c^2} \int_0^c r f(r)\, dr$$

$$A_n = \frac{2}{c^2 J_0^2(\lambda_n c)} \int_0^c r J_0(\lambda_n r) f(r)\, dr.$$

9. Substituting $u(r,t) = v(r,t) + \psi(r)$ into the partial differential equation gives

$$\frac{\partial^2 v}{\partial r^2} + \frac{1}{r}\frac{\partial v}{\partial r} + \psi'' + \frac{1}{r}\psi' = \frac{\partial v}{\partial t}.$$

This equation will be homogeneous provided $\psi'' + \frac{1}{r}\psi' = 0$ or

$$\psi(r) = c_1 \ln r + c_2.$$

Since $\ln r$ is unbounded as $r \to 0$ we take $c_1 = 0$. Then $\psi(r) = c_2$ and using

$$u(2,t) = v(2,t) + \psi(2) = 100$$

we set $c_2 = \psi(r) = 100$. Referring to Problem 6 above, the solution of

$$\frac{\partial^2 v}{\partial r^2} + \frac{1}{r}\frac{\partial v}{\partial r} = \frac{\partial v}{\partial t}, \qquad 0 < r < 2, \quad t > 0$$

is

$$v(r,t) = c_1 J_0(\lambda r) e^{\mu t}.$$

The boundary conditions

$$v(2,t) = 0, \quad t > 0,$$

$$v(r,0) = u(r,0) - \psi(r)$$

then give

$$v(r,t) = \sum_{n=1}^{\infty} A_n J_0(\lambda_n r) e^{-\lambda_n^2 t}$$

where

$$\begin{aligned}
A_n &= \frac{2}{2^2 J_1^2(2\lambda_n)} \int_0^2 r J_0(\lambda_n r)[u(r,0) - \psi(r)]\,dr \\
&= \frac{1}{2J_1^2(2\lambda_n)} \left[\int_0^1 r J_0(\lambda_n r)[200 - 100]\,dr + \int_1^2 r J_0(\lambda_n r)[100 - 100]\,dr \right] \\
&= \frac{50}{J_1^2(2\lambda_n)} \int_0^1 r J_0(\lambda_n r)\,dr \qquad \boxed{x = \lambda_n r,\ dx = \lambda_n\,dr} \\
&= \frac{50}{J_1^2(2\lambda_n)} \int_0^{\lambda_n} \frac{1}{\lambda_n^2} x J_0(x)\,dx \\
&= \frac{50}{\lambda_n^2 J_1^2(2\lambda_n)} \int_0^{\lambda_n} \frac{d}{dx}[xJ_1(x)]\,dx \qquad \boxed{\text{see (4) of Section 11.5 in text}} \\
&= \frac{50}{\lambda_n^2 J_1^2(2\lambda_n)} (xJ_1(x)) \Big|_0^{\lambda_n} = \frac{50 J_1(\lambda_n)}{\lambda_n J_1^2(2\lambda_n)}.
\end{aligned}$$

Thus

$$u(r,t) = v(r,t) + \psi(r) = 100 + 50 \sum_{n=1}^{\infty} \frac{J_1(\lambda_n) J_0(\lambda_n r)}{\lambda_n J_1^2(2\lambda_n)} e^{-\lambda^2 t}.$$

12. (a) First we see that

$$\frac{R''\Theta + \frac{1}{r} R'\Theta + \frac{1}{r^2} R\Theta''}{R\Theta} = \frac{T''}{a^2 T} = -\lambda^2.$$

This gives $T'' + a^2\lambda^2 T = 0$. Then from

$$\frac{R'' + \frac{1}{r} R' + \lambda^2 R}{-R/r^2} = \frac{\Theta''}{\Theta} = -\nu^2$$

we get $\Theta'' + \nu^2\Theta = 0$ and $r^2 R'' + rR' + (\lambda^2 r^2 - \nu^2)R = 0$.

(b) The general solutions of the differential equations in part (a) are

$$T = c_1 \cos a\lambda t + c_2 \sin a\lambda t$$

$$\Theta = c_3 \cos \nu\theta + c_4 \cos \nu\theta$$

$$R = c_5 J_\nu(\lambda r) + c_6 Y_\nu(\lambda r).$$

(c) Implicitly we expect $u(r,\theta,t) = u(r,\theta + 2\pi, t)$ and so Θ must be 2π-periodic. Therefore $\nu = n$, $n = 0, 1, 2, \ldots$. The corresponding eigenfunctions are 1, $\cos\theta$, $\cos 2\theta$, $\ldots$, $\sin\theta$, $\sin 2\theta$, $\ldots$.

Arguing that $u(r,\theta,t)$ is bounded as $r \to 0$ we then define $c_6 = 0$ and so $R = c_3 J_n(\lambda r)$. But $R(c) = 0$ gives $J_n(\lambda c) = 0$; this equation defines the eigenvalues λ_n. For each n, $\lambda_{ni} = x_{ni}/c$, $i = 1, 2, 3, \ldots$.

(d) $u(r,\theta,t) = \sum_{i=1}^{n}(A_{0i}\cos a\lambda_{0i}t + B_{0i}\sin a\lambda_{0i}t)J_0(\lambda_{0i}r)$

$$+\sum_{n=1}^{\infty}\sum_{i=1}^{\infty}\Big[(A_{ni}\cos a\lambda_{ni}t + B_{ni}\sin a\lambda_{ni}t)\cos n\theta$$

$$+ (C_{ni}\cos a\lambda_{ni}t + D_{ni}\sin a\lambda_{ni}t)\sin n\theta\Big]J_n(\lambda_{ni}r)$$

Exercises 13.3

3. The coefficients are given by

$$A_n = \frac{2n+1}{2c^n}\int_0^{\pi}\cos\theta P_n(\cos\theta)\sin\theta\,d\theta = \frac{2n+1}{2c^n}\int_0^{\pi}P_1(\cos\theta)P_n(\cos\theta)\sin\theta\,d\theta$$

$$\boxed{x = \cos\theta,\ dx = -\sin\theta\,d\theta}$$

$$= \frac{2n+1}{2c^n}\int_{-1}^{1}P_1(x)P_n(x)\,dx.$$

Since $P_n(x)$ and $P_m(x)$ are orthogonal for $m \neq n$, $A_n = 0$ for $n \neq 1$ and

$$A_1 = \frac{2(1)+1}{2c^1}\int_{-1}^{1}P_1(x)P_1(x)\,dx = \frac{3}{2c}\int_{-1}^{1}x^2dx = \frac{1}{c}.$$

Thus

$$u(r,\theta) = \frac{r}{c}P_1(\cos\theta) = \frac{r}{c}\cos\theta.$$

6. Referring to Example 1 in the text we have

$$R(r) = c_1 r^n \quad \text{and} \quad \Theta(\theta) = P_n(\cos\theta).$$

Now $\Theta(\pi/2) = 0$ implies that n is odd, so

$$u(r,\theta) = \sum_{n=0}^{\infty}A_{2n+1}r^{2n+1}P_{2n+1}(\cos\theta).$$

From

$$u(c,\theta) = f(\theta) = \sum_{n=0}^{\infty}A_{2n+1}c^{2n+1}P_{2n+1}(\cos\theta)$$

we see that

$$A_{2n+1}c^{2n+1} = (4n+3)\int_0^{\pi/2}f(\theta)\sin\theta\,P_{2n+1}(\cos\theta)\,d\theta.$$

Thus

$$u(r,\theta)=\sum_{n=0}^{\infty}A_{2n+1}r^{2n+1}P_{2n+1}(\cos\theta)$$

where

$$A_{2n+1}=\frac{4n+3}{c^{2n+1}}\int_0^{\pi/2}f(\theta)\sin\theta\,P_{2n+1}(\cos\theta)\,d\theta.$$

9. Checking the hint, we find

$$\frac{1}{r}\frac{\partial^2}{\partial r^2}(ru)=\frac{1}{r}\frac{\partial}{\partial r}\left[r\frac{\partial u}{\partial r}+u\right]=\frac{1}{r}\left[r\frac{\partial^2 u}{\partial r^2}+\frac{\partial u}{\partial r}+\frac{\partial u}{\partial r}\right]=\frac{\partial^2 u}{\partial r^2}+\frac{2}{r}\frac{\partial u}{\partial r}.$$

The partial differential equation then becomes

$$\frac{\partial^2}{\partial r^2}(ru)=r\frac{\partial u}{\partial t}.$$

Now, letting $ru(r,t)=v(r,t)+\psi(r)$, since the boundary condition is nonhomogeneous, we obtain

$$\frac{\partial^2}{\partial r^2}\left[v(r,t)+\psi(r)\right]=r\frac{\partial}{\partial t}\left[\frac{1}{r}v(r,t)+\psi(r)\right]$$

or

$$\frac{\partial^2 v}{\partial r^2}+\psi''(r)=\frac{\partial v}{\partial t}.$$

This differential equation will be homogeneous if $\psi''(r)=0$ or $\psi(r)=c_1r+c_2$. Now

$$u(r,t)=\frac{1}{r}v(r,t)+\frac{1}{r}\psi(r)\quad\text{and}\quad\frac{1}{r}\psi(r)=c_1+\frac{c_2}{r}.$$

Since we want $u(r,t)$ to be bounded as r approaches 0, we require $c_2=0$. Then $\psi(r)=c_1r$. When $r=1$

$$u(1,t)=v(1,t)+\psi(1)=v(1,t)+c_1=100,$$

and we will have the homogeneous boundary condition $v(1,t)=0$ when $c_1=100$. Consequently, $\psi(r)=100r$. The initial condition

$$u(r,0)=\frac{1}{r}v(r,0)+\frac{1}{r}\psi(r)=\frac{1}{r}v(r,0)+100=0$$

implies $v(r,0)=-100r$. We are thus led to solve the new boundary-value problem

$$\frac{\partial^2 v}{\partial r^2}=\frac{\partial v}{\partial t},\quad 0<r<1,\quad t>0,$$

$$v(1,t)=0,\quad \lim_{r\to 0}\frac{1}{r}v(r,t)<\infty,$$

$$v(r,0)=-100r.$$

Letting $v(r,t) = R(r)T(t)$ and separating variables leads to

$$R'' + \lambda^2 R = 0 \quad \text{and} \quad T' + \lambda^2 T = 0$$

with solutions

$$R(r) = c_3 \cos \lambda r + c_4 \sin \lambda r \quad \text{and} \quad T(t) = c_5 e^{-\lambda^2 t}.$$

The boundary conditions are equivalent to $R(1) = 0$ and $\lim_{r\to 0} \frac{1}{r} R(r) < \infty$. Since

$$\lim_{r\to 0} \frac{1}{r} R(r) = \lim_{r\to 0} \frac{c_3 \cos \lambda r}{r} + \lim_{r\to 0} \frac{c_4 \sin \lambda r}{r} = \lim_{r\to 0} \frac{c_3 \cos \lambda r}{r} + c_4 \lambda < \infty$$

we must have $c_3 = 0$. Then $R(r) = c_4 \sin \lambda r$, and $R(1) = 0$ implies $\lambda = n\pi$ for $n = 1, 2, 3, \ldots$. Thus

$$v_n(r,t) = A_n e^{-n^2\pi^2 t} \sin n\pi r$$

for $n = 1, 2, 3, \ldots$. Using the condition $\lim_{r\to 0} \frac{1}{r} R(r) < \infty$ it is easily shown that there are no eigenvalues for $\lambda = 0$, nor does setting the common constant to $+\lambda^2$ when separating variables lead to any solutions. Now, by the superposition principle,

$$v(r,t) = \sum_{n=1}^{\infty} A_n e^{-n^2\pi^2 t} \sin n\pi r.$$

The initial condition $v(r,0) = -100r$ implies

$$-100r = \sum_{n=1}^{\infty} A_n \sin n\pi r.$$

This is a Fourier sine series and so

$$A_n = 2\int_0^1 (-100r \sin n\pi r)\, dr = -200 \left[-\frac{r}{n\pi} \cos n\pi r \Big|_0^1 + \int_0^1 \frac{1}{n\pi} \cos n\pi r \, dr \right]$$

$$= -200 \left[-\frac{\cos n\pi}{n\pi} + \frac{1}{n^2\pi^2} \sin n\pi r \Big|_0^1 \right] = -200 \left[-\frac{(-1)^n}{n\pi} \right] = \frac{(-1)^n 200}{n\pi}.$$

A solution of the problem is thus

$$u(r,t) = \frac{1}{r} v(r,t) + \frac{1}{r}\psi(r) = \frac{1}{r} \sum_{n=1}^{\infty} (-1)^n \frac{20}{n\pi} e^{-n^2\pi^2 t} \sin n\pi r + \frac{1}{r}(100r)$$

$$= \frac{200}{\pi r} \sum_{n=1}^{\infty} \frac{(-1)^n}{n} e^{-n^2\pi^2 t} \sin n\pi r + 100.$$

12. Proceeding as in Example 1 we obtain

$$\Theta(\theta) = P_n(\cos\theta) \quad \text{and} \quad R(r) = c_1 r^n + c_2 r^{-(n+1)}$$

so that

$$u(r,\theta) = \sum_{n=0}^{\infty} (A_n r^n + B_n r^{-(n+1)}) P_n(\cos\theta).$$

To satisfy $\lim_{r\to\infty} u(r,\theta) = -Er\cos\theta$ we must have $A_n = 0$ for $n = 2, 3, 4, \ldots$. Then

$$\lim_{r\to\infty} u(r,\theta) = -Er\cos\theta = A_0 \cdot 1 + A_1 r\cos\theta,$$

so $A_0 = 0$ and $A_1 = -E$. Thus

$$u(r,\theta) = -Er\cos\theta + \sum_{n=0}^{\infty} B_n r^{-(n+1)} P_n(\cos\theta).$$

Now

$$u(c,\theta) = 0 = -Ec\cos\theta + \sum_{n=0}^{\infty} B_n c^{-(n+1)} P_n(\cos\theta)$$

so

$$\sum_{n=0}^{\infty} B_n c^{-(n+1)} P_n(\cos\theta) = Ec\cos\theta$$

and

$$B_n c^{-(n+1)} = \frac{2n+1}{2} \int_0^{\pi} Ec\cos\theta\, P_n(\cos\theta)\sin\theta\, d\theta.$$

Now $\cos\theta = P_1(\cos\theta)$ so, for $n \neq 1$,

$$\int_0^{\pi} \cos\theta\, P_n(\cos\theta)\sin\theta\, d\theta = 0$$

by orthogonality. Thus $B_n = 0$ for $n \neq 1$ and

$$B_1 = \frac{3}{2} Ec^3 \int_0^{\pi} \cos^2\theta \sin\theta\, d\theta = Ec^3.$$

Therefore,

$$u(r,\theta) = -Er\cos\theta + Ec^3 r^{-2}\cos\theta.$$

Chapter 13 Review Exercises

3. The conditions $\Theta(0) = 0$ and $\Theta(\pi) = 0$ applied to $\Theta = c_1\cos\lambda\theta + c_2\sin\lambda\theta$ give $c_1 = 0$ and $\lambda = n$, $n = 1, 2, 3, \ldots$, respectively. Thus we have the Fourier sine-series coefficients

$$A_n = \frac{2}{\pi} \int_0^{\pi} u_0(\pi\theta - \theta^2)\sin n\theta\, d\theta = \frac{4u_0}{n^3\pi}[1 - (-1)^n].$$

Thus

$$u(r,\theta) = \frac{4u_0}{\pi} \sum_{n=1}^{\infty} \frac{1-(-1)^n}{n^3} r^n \sin n\theta.$$

6. We solve

$$\frac{\partial^2 u}{\partial r^2} + \frac{1}{r}\frac{\partial u}{\partial r} + \frac{1}{r^2}\frac{\partial^2 u}{\partial \theta^2} = 0, \quad r > 1, \quad 0 < \theta < \pi,$$

$$u(r,0) = 0, \quad u(r,\pi) = 0, \quad r > 1,$$

$$u(1,\theta) = f(\theta), \quad 0 < \theta < \pi.$$

Separating variables we obtain

$$\Theta(\theta) = c_1 \cos \lambda\theta + c_2 \sin \lambda\theta$$

$$R(r) = c_3 r^{\lambda} + c_4 r^{-\lambda}.$$

Applying the boundary conditions $\Theta(0) = 0$, and $\Theta(\pi) = 0$ gives $c_1 = 0$ and $\lambda = n$ for $n = 1, 2, 3, \ldots$. Assuming $f(\theta)$ to be bounded, we expect the solution $u(r,\theta)$ to also be bounded as $r \to \infty$. This requires that $c_3 = 0$. Therefore

$$u(r,\theta) = \sum_{n=1}^{\infty} A_n r^{-n} \sin n\theta.$$

From

$$u(1,\theta) = f(\theta) = \sum_{n=1}^{\infty} A_n \sin n\theta$$

we obtain

$$A_n = \frac{2}{\pi}\int_0^{\pi} f(\theta) \sin n\theta \, d\theta.$$

9. Referring to Example 2 in Section 13.2 we have

$$R(r) = c_1 J_0(\lambda r) + c_2 Y_0(\lambda r)$$

$$Z(z) = c_3 \cosh \lambda z + c_4 \sinh \lambda z$$

where $c_2 = 0$ and $J_0(2\lambda) = 0$ defines the positive eigenvalues λ_n. From $Z'(0) = 0$ we obtain $c_4 = 0$. Then

$$u(r,z) = \sum_{n=1}^{\infty} A_n \cosh \lambda_n z J_0(\lambda_n r).$$

From

$$u(r,4) = 50 = \sum_{n=1}^{\infty} A_n \cosh 4\lambda_n J_0(\lambda_n r)$$

we obtain (as in Example 1 of Section 13.1)

$$A_n \cosh 4\lambda_n = \frac{2(50)}{4J_1^2(2\lambda_n)}\int_0^2 r J_0(\lambda_n r)\, dr = \frac{50}{\lambda_n J_1(2\lambda_n)}.$$

Thus the temperature in the cylinder is

$$u(r,z) = 50\sum_{n=1}^{\infty} \frac{\cosh \lambda_n z J_0(\lambda_n r)}{\lambda_n \cosh 4\lambda_n J_1(2\lambda_n)}.$$

12. Since

$$\frac{1}{r}\frac{\partial^2}{\partial r^2}(ru) = \frac{1}{r}\frac{\partial}{\partial r}\left[r\frac{\partial u}{\partial r} + u\right] = \frac{1}{r}\left[r\frac{\partial^2 u}{\partial r^2} + \frac{\partial u}{\partial r} + \frac{\partial u}{\partial r}\right] = \frac{\partial^2 u}{\partial r^2} + \frac{2}{r}\frac{\partial u}{r}$$

the differential equation becomes

$$\frac{1}{r}\frac{\partial^2}{\partial r^2}(ru) = \frac{\partial^2 u}{\partial t^2} \quad \text{or} \quad \frac{\partial^2}{\partial r^2}(ru) = r\frac{\partial^2 u}{\partial t^2}.$$

Letting $v(r,t) = ru(r,t)$ we obtain the boundary-value problem

$$\frac{\partial^2 v}{\partial r^2} = \frac{\partial^2 v}{\partial t^2}, \quad 0 < r < 1, \quad t > 0$$

$$\left.\frac{\partial v}{\partial r}\right|_{r=1} - v(1,t) = 0, \quad t > 0$$

$$v(r,0) = rf(r), \quad \left.\frac{\partial v}{\partial t}\right|_{t=0} = rg(r), \quad 0 < r < 1.$$

If we separate variables using $v(r,t) = R(r)T(t)$ then we obtain

$$R(r) = c_1\cos\lambda r + c_2\sin\lambda r$$

$$T(t) = c_3\cos\lambda t + c_4\sin\lambda t.$$

Since $u(r,t) = v(r,t)/r$, in order to insure boundedness at $r = 0$ we define $c_1 = 0$. Then $R(r) = c_2\sin\lambda r$. Now the boundary condition $R'(1) - R(1) = 0$ implies $\lambda\cos\lambda - \sin\lambda = 0$. Thus, the eigenvalues λ_n are the positive solutions of $\tan\lambda = \lambda$. We now have

$$v_n(r,t) = (A_n\cos\lambda_n t + B_n\sin\lambda_n t)\sin\lambda_n r.$$

For the eigenvalue $\lambda = 0$,

$$R(r) = c_1 r + c_2 \quad \text{and} \quad T(t) = c_3 t + c_4,$$

and boundedness at $r = 0$ implies $c_2 = 0$. We then take

$$v_0(r,t) = A_0 tr + B_0 r$$

so that

$$v(r,t) = A_0 tr + B_0 r + \sum_{n=1}^{\infty}(a_n\cos\lambda_n t + B_n\sin\lambda_n t)\sin\lambda_n r.$$

Now

$$v(r,0) = rf(r) = B_0 r + \sum_{n=1}^{\infty} A_n\sin\lambda_n r.$$

Since $\{r, \sin\lambda_n r\}$ is an orthogonal set on $[0,1]$,

$$\int_0^1 r\sin\lambda_n r\,dr = 0 \quad \text{and} \quad \int_0^1 \sin\lambda_n r \sin\lambda_n r\,dr = 0$$

for $m \neq n$. Therefore

$$\int_0^1 r^2 f(r)\,dr = B_0 \int_0^1 r^2\,dr = \frac{1}{3}B_0$$

and

$$B_0 = 3\int_0^1 r^2 f(r)\,dr.$$

Also

$$\int_0^1 r f(r)\sin\lambda_n r\,dr = A_n \int_0^1 \sin^2\lambda_n r\,dr$$

and

$$A_n = \frac{\int_0^1 r f(r)\sin\lambda_n r\,dr}{\int_0^1 \sin^2\lambda_n r\,dr}.$$

Now

$$\int_0^1 \sin^2\lambda_n r\,dr = \frac{1}{2}\int_0^1 (1-\cos 2\lambda_n r)\,dr = \frac{1}{2}\left[1 - \frac{\sin 2\lambda_n}{2\lambda_n}\right] = \frac{1}{2}[1-\cos^2\lambda_n].$$

Since $\tan\lambda_n = \lambda_n$,

$$1+\lambda_n^2 = 1+\tan^2\lambda_n = \sec^2\lambda_n = \frac{1}{\cos^2\lambda_n}$$

and

$$\cos^2\lambda_n = \frac{1}{1+\lambda_n^2}.$$

Then

$$\int_0^1 \sin^2\lambda_n r\,dr = \frac{1}{2}\left[1 - \frac{1}{1+\lambda_n^2}\right] = \frac{\lambda_n^2}{2(1+\lambda_n^2)}$$

and

$$A_n = \frac{2(1+\lambda_n^2)}{\lambda_n^2}\int_0^1 r f(r)\sin\lambda_n r\,dr.$$

Similarly, setting

$$\left.\frac{\partial v}{\partial t}\right|_{t=0} = r g(r) = A_0 r + \sum_{n=1}^{\infty} B_n \lambda_n \sin\lambda_n r$$

we obtain

$$A_0 = 3\int_0^1 r^2 g(r)\,dr$$

and

$$B_n = \frac{2(1+\lambda_n^2)}{\lambda_n^3}\int_0^1 r g(r)\sin\lambda_n r\,dr.$$

Therefore, since $v(r,t) = ru(r,t)$ we have

$$u(r,t) = A_0 t + B_0 + \sum_{n=1}^{\infty} (A_n \cos \lambda_n t + B_n \sin \lambda_n t) \frac{\sin \lambda_n r}{r},$$

where the λ_n are solutions of $\tan \lambda = \lambda$ and

$$A_0 = 3 \int_0^1 r^2 g(r)\, dr$$

$$B_0 = 3 \int_0^1 r^2 f(r)\, dr$$

$$A_n = \frac{2(1+\lambda_n^2)}{\lambda_n^2} \int_0^1 r f(r) \sin \lambda_n r\, dr$$

$$B_n = \frac{2(1+\lambda_n^2)}{\lambda_n^3} \int_0^1 r g(r) \sin \lambda_n r\, dr$$

for $n = 1, 2, 3, \ldots$.

14 Integral Transform Method

Exercises 14.1

3. By the first translation theorem,

$$\mathscr{L}\left\{e^t \operatorname{erf}(\sqrt{t})\right\} = \mathscr{L}\left\{\operatorname{erf}(\sqrt{t})\right\}\Big|_{s\to s-1} = \frac{1}{s\sqrt{s+1}}\Big|_{s\to s-1} = \frac{1}{\sqrt{s}\,(s-1)}.$$

6. We first compute

$$\frac{\sinh a\sqrt{s}}{s\sinh\sqrt{s}} = \frac{e^{a\sqrt{s}} - e^{-a\sqrt{s}}}{s(e^{\sqrt{s}} - e^{-\sqrt{s}})} = \frac{e^{(a-1)\sqrt{s}} - e^{-(a+1)\sqrt{s}}}{s(1 - e^{-2\sqrt{s}})}$$

$$= \frac{e^{(a-1)\sqrt{s}}}{s}\left[1 + e^{-2\sqrt{s}} + e^{-4\sqrt{s}} + \cdots\right] - \frac{e^{-(a+1)\sqrt{s}}}{s}\left[1 + e^{-2\sqrt{s}} + e^{-4\sqrt{s}} + \cdots\right]$$

$$= \left[\frac{e^{-(1-a)\sqrt{s}}}{s} + \frac{e^{-(3-a)\sqrt{s}}}{s} + \frac{e^{-(5-a)\sqrt{s}}}{s} + \cdots\right]$$

$$- \left[\frac{e^{-(1+a)\sqrt{s}}}{s} + \frac{e^{-(3+a)\sqrt{s}}}{s} + \frac{e^{-(5+a)\sqrt{s}}}{s} + \cdots\right]$$

$$= \sum_{n=0}^{\infty}\left[\frac{e^{-(2n+1-a)\sqrt{s}}}{s} - \frac{e^{-(2n+1+a)\sqrt{s}}}{s}\right].$$

Then

$$\mathscr{L}\left\{\frac{\sinh a\sqrt{s}}{s\sinh\sqrt{s}}\right\} = \sum_{n=0}^{\infty}\left[\mathscr{L}\left\{\frac{e^{-(2n+1-a)\sqrt{s}}}{s}\right\} - \mathscr{L}\left\{-\frac{e^{-(2n+1+a)\sqrt{s}}}{s}\right\}\right]$$

$$= \sum_{n=0}^{\infty}\left[\operatorname{erfc}\left(\frac{2n+1-a}{2\sqrt{t}}\right) - \operatorname{erfc}\left(\frac{2n+1+a}{2\sqrt{t}}\right)\right]$$

$$= \sum_{n=0}^{\infty}\left(\left[1 - \operatorname{erf}\left(\frac{2n+1-a}{2\sqrt{t}}\right)\right] - \left[1 - \operatorname{erf}\left(\frac{2n+1+a}{2\sqrt{t}}\right)\right]\right)$$

$$= \sum_{n=0}^{\infty}\left[\operatorname{erf}\left(\frac{2n+1+a}{2\sqrt{t}}\right) - \operatorname{erf}\left(\frac{2n+1-a}{2\sqrt{t}}\right)\right].$$

9. $\displaystyle\int_a^b e^{-u^2}\,du = \int_a^0 e^{-u^2}\,du + \int_0^b e^{-u^2}\,du = \int_0^b e^{-u^2}\,du - \int_0^a e^{-u^2}\,du$

$$= \frac{\sqrt{\pi}}{2}\operatorname{erf}(b) - \frac{\sqrt{\pi}}{2}\operatorname{erf}(a) = \frac{\sqrt{\pi}}{2}[\operatorname{erf}(b) - \operatorname{erf}(a)]$$

Exercises 14.2

3. The solution of

$$a^2\frac{d^2U}{dx^2}-s^2U=0$$

is in this case

$$U(x,s)=c_1e^{-(x/a)s}+c_2e^{(x/a)s}.$$

Since $\lim_{x\to\infty}u(x,t)=0$ we have $\lim_{x\to\infty}U(x,s)=0$. Thus $c_2=0$ and

$$U(x,s)=c_1e^{-(x/a)s}.$$

If $\mathscr{L}\{u(0,t)\}=\mathscr{L}\{f(t)\}=F(s)$ then $U(0,s)=F(s)$. From this we have $c_1=F(s)$ and

$$U(x,s)=F(s)e^{-(x/a)s}.$$

Hence, by the second translation theorem,

$$u(x,t)=f\left(t-\frac{x}{a}\right)\mathscr{U}\left(t-\frac{x}{a}\right).$$

6. Transforming the partial differential equation gives

$$\frac{d^2U}{dx^2}-s^2U=-\frac{\omega}{s^2+\omega^2}\sin\pi x.$$

Using undetermined coefficients we obtain

$$U(x,s)=c_1\cosh sx+c_2\sinh sx+\frac{\omega}{(s^2+\pi^2)(s^2+\omega^2)}\sin\pi x.$$

The transformed boundary conditions $U(0,s)=0$ and $U(1,s)=0$ give, in turn, $c_1=0$ and $c_2=0$. Therefore

$$U(x,s)=\frac{\omega}{(s^2+\pi^2)(s^2+\omega^2)}\sin\pi x$$

and

$$\begin{aligned}
u(x,t)&=\omega\sin\pi x\,\mathscr{L}^{-1}\left\{\frac{1}{(s^2+\pi^2)(s^2+\omega^2)}\right\}\\
&=\frac{\omega}{\omega^2-\pi^2}\sin\pi x\,\mathscr{L}^{-1}\left\{\frac{1}{\pi}\frac{\pi}{s^2+\pi^2}-\frac{1}{\omega}\frac{\omega}{s^2+\omega^2}\right\}\\
&=\frac{\omega}{\pi(\omega^2-\pi^2)}\sin\pi t\sin\pi x-\frac{1}{\omega^2-\pi^2}\sin\omega t\sin\pi x.
\end{aligned}$$

9. Transforming the partial differential equation gives

$$\frac{d^2U}{dx^2}-s^2U=-sxe^{-x}.$$

Using undetermined coefficients we obtain

$$U(x,s) = c_1 e^{-sx} + c_2 e^{sx} - \frac{2s}{(s^2-1)^2}\, e^{-x} + \frac{s}{s^2-1}\, xe^{-x}.$$

The transformed boundary conditions $\lim_{x\to\infty} U(x,s) = 0$ and $U(0,s) = 0$ give, in turn, $c_2 = 0$ and $c_1 = 2s/(s^2-1)^2$. Therefore

$$U(x,s) = \frac{2s}{(s^2-1)^2}\, e^{-sx} - \frac{2s}{(s^2-1)^2}\, e^{-x} + \frac{s}{s^2-1}\, xe^{-x}.$$

From entries (13) and (26) in the table we obtain

$$\begin{aligned} u(x,t) &= \mathscr{L}^{-1}\left\{\frac{2s}{(s^2-1)^2}\, e^{-sx} - \frac{2s}{(s^2-1)^2}\, e^{-x} + \frac{s}{s^2-1}\, xe^{-x}\right\} \\ &= 2(t-x)\sinh(t-x)\,\mathscr{U}(t-x) - te^{-x}\sinh t + xe^{-x}\cosh t. \end{aligned}$$

12. Transforming the partial differential equation and using the initial condition gives

$$k\,\frac{d^2U}{dx^2} - sU = 0.$$

Since the domain of the variable x is an infinite interval we write the general solution of this differential equation as

$$U(x,s) = c_1 e^{-\sqrt{s/k}\,x} + c_2 e^{-\sqrt{s/k}\,x}.$$

Transforming the boundary conditions gives $U'(0,s) = -A/s$ and $\lim\limits_{x\to\infty} U(x,s) = 0$. Hence we find $c_2 = 0$ and $c_1 = A\sqrt{k}/s\sqrt{s}$. From

$$U(x,s) = A\sqrt{k}\,\frac{e^{-\sqrt{s/k}\,x}}{s\sqrt{s}}$$

we see that

$$u(x,t) = A\sqrt{k}\,\mathscr{L}^{-1}\left\{\frac{e^{-\sqrt{s/k}\,x}}{s\sqrt{s}}\right\}.$$

With the identification $a = x/\sqrt{k}$ it follows from entry 55 of the table in Appendix III that

$$\begin{aligned} u(x,t) &= A\sqrt{k}\left\{2\sqrt{\frac{t}{\pi}}\, e^{-x^2/4kt} - \frac{x}{\sqrt{k}}\operatorname{erfc}\left(x/2\sqrt{kt}\right)\right\} \\ &= 2A\sqrt{\frac{kt}{\pi}}\, e^{-x^2/4kt} - Ax\operatorname{erfc}\left(x/2\sqrt{kt}\right). \end{aligned}$$

15. We use

$$U(x,s) = c_1 e^{-\sqrt{s}\,x} + c_2 e^{\sqrt{s}\,x} + \frac{u_0}{s}.$$

The condition $\lim_{x\to\infty} u(x,t) = u_0$ implies $\lim_{x\to\infty} U(x,s) = u_0/s$, so we define $c_2 = 0$. Then

$$U(x,s) = c_1 e^{-\sqrt{s}\,x} + \frac{u_0}{s}.$$

The transform of the remaining boundary conditions gives

$$\left.\frac{dU}{dx}\right|_{x=0} = U(0,s).$$

This condition yields $c_1 = -u_0/s(\sqrt{s}+1)$. Thus

$$U(x,s) = -u_0\,\frac{e^{-\sqrt{s}\,x}}{s(\sqrt{s}+1)} + \frac{u_0}{s}$$

and

$$u(x,t) = -u_0\,\mathscr{L}^{-1}\left\{\frac{e^{-x\sqrt{s}}}{s(\sqrt{s}+1)}\right\} + u_0\,\mathscr{L}^{-1}\left\{\frac{1}{s}\right\}$$

$$= u_0 e^{x+t}\operatorname{erfc}\left(\sqrt{t} + \frac{x}{2\sqrt{t}}\right) - u_0\operatorname{erfc}\left(\frac{x}{2\sqrt{t}}\right) + u_0 \qquad \boxed{\text{By (5) in the table in 14.1.}}$$

18. We use

$$U(x,s) = c_1 e^{-\sqrt{s}\,x} + c_2 e^{\sqrt{s}\,x}.$$

The condition $\lim_{x\to\infty} u(x,t) = 0$ implies $\lim_{x\to\infty} U(x,s) = 0$, so we define $c_2 = 0$. Then $U(x,s) = c_1 e^{-\sqrt{s}\,x}$. The transform of the remaining boundary condition gives

$$\left.\frac{dU}{dx}\right|_{x=0} = -F(s)$$

where $F(s) = \mathscr{L}\{f(t)\}$. This condition yields $c_1 = F(s)/\sqrt{s}$. Thus

$$U(x,s) = F(s)\,\frac{e^{-\sqrt{s}\,x}}{\sqrt{s}}.$$

Using entry (44) of the table and the convolution theorem we obtain

$$u(x,t) = \mathscr{L}^{-1}\left\{F(s)\cdot\frac{e^{-\sqrt{s}\,x}}{\sqrt{s}}\right\} = \frac{1}{\sqrt{\pi}}\int_0^t f(\tau)\,\frac{e^{-x^2/4(t-\tau)}}{\sqrt{t-\tau}}\,d\tau.$$

21. Transforming the partial differential equation gives

$$\frac{d^2U}{dx^2} - sU = 0$$

and so

$$U(x,s) = c_1 e^{-\sqrt{s}\,x} + c_2 e^{\sqrt{s}\,x}.$$

The condition $\lim_{x\to-\infty} u(x,t) = 0$ implies $\lim_{x\to-\infty} U(x,s) = 0$, so we define $c_1 = 0$. The transform of the remaining boundary condition gives

$$\left.\frac{dU}{dx}\right|_{x=1} = \frac{100}{s} - U(1,s).$$

This condition yields

$$c_2\sqrt{s}\,e^{\sqrt{s}} = \frac{100}{s} - c_2 e^{\sqrt{s}}$$

from which it follows that

$$c_2 = \frac{100}{s(\sqrt{s}+1)}\,e^{-\sqrt{s}}.$$

Thus

$$U(x,s) = 100\,\frac{e^{-(1-x)\sqrt{s}}}{s(\sqrt{s}+1)}.$$

Using entry (57) of the table in Appendix III we obtain

$$u(x,t) = 100\,\mathscr{L}^{-1}\left\{\frac{e^{-(1-x)\sqrt{s}}}{s(\sqrt{s}+1)}\right\} = 100\left[-e^{1-x+t}\operatorname{erfc}\left(\sqrt{t}+\frac{1-x}{\sqrt{t}}\right) + \operatorname{erfc}\left(\frac{1-x}{2\sqrt{t}}\right)\right].$$

24. The transform of the partial differential equation is

$$k\frac{d^2U}{dx^2} - hU + h\frac{u_m}{s} = sU - u_0$$

or

$$k\frac{d^2U}{dx^2} - (h+s)U = -h\frac{u_m}{s} - u_0.$$

By undetermined coefficients we find

$$U(x,s) = c_1 e^{\sqrt{(h+s)/k}\,x} + c_2 e^{-\sqrt{(h+s)/k}\,x} + \frac{hu_m + u_0 s}{s(s+h)}.$$

The transformed boundary conditions are $U'(0,s) = 0$ and $U'(L,s) = 0$. These conditions imply $c_1 = 0$ and $c_2 = 0$. By partial fractions we then get

$$U(x,s) = \frac{hu_m + u_0 s}{s(s+h)} = \frac{u_m}{s} - \frac{u_m}{s+h} + \frac{u_0}{s+h}.$$

Therefore,

$$u(x,t) = u_m\,\mathscr{L}^{-1}\left\{\frac{1}{s}\right\} - u_m\,\mathscr{L}^{-1}\left\{\frac{1}{s+h}\right\} + u_0\,\mathscr{L}^{-1}\left\{\frac{1}{s+h}\right\} = u_m - u_m e^{-ht} + u_0 e^{-ht}.$$

27. We use

$$U(x,s) = c_1 e^{-\sqrt{RCs+RG}\,x} + c_2 e^{\sqrt{RCs+RG}} + \frac{Cu_0}{Cs+G}.$$

The condition $\lim_{x\to\infty} \partial u/\partial x = 0$ implies $\lim_{x\to\infty} dU/dx = 0$, so we define $c_2 = 0$. Applying $U(0,s) = 0$ to

$$U(x,s) = c_1 e^{-\sqrt{RCsRG}\,x} + \frac{Cu_0}{Cs+G}$$

gives $c_1 = -Cu_0/(Cs+G)$. Therefore

$$U(x,s) = -Cu_0 \frac{e^{-\sqrt{RCs+RG}\,x}}{Cs+G} + \frac{Cu_0}{Cs+G}$$

and

$$\begin{aligned}
u(x,t) &= u_0 \mathscr{L}^{-1}\left\{\frac{1}{s+G/C}\right\} - u_0 \mathscr{L}^{-1}\left\{\frac{e^{-x\sqrt{RC}\sqrt{s+G/C}}}{s+G/C}\right\} \\
&= u_0 e^{-Gt/C} - u_0 e^{-Gt/C} \operatorname{erfc}\left(\frac{x\sqrt{RC}}{2\sqrt{t}}\right) \\
&= u_0 e^{-Gt/C}\left[1 - \operatorname{erfc}\left(\frac{x}{2}\sqrt{\frac{RC}{t}}\right)\right] \\
&= u_0 e^{-Gt/C} \operatorname{erf}\left(\frac{x}{2}\sqrt{\frac{RC}{t}}\right).
\end{aligned}$$

30. (a) We use

$$U(x,s) = c_1 e^{-(s/a)x} + c_2 e^{(s/a)x} + \frac{v_0^2 F_0}{(a^2 - v_0^2)s^2} e^{-(s/v_0)x}.$$

The condition $\lim_{x\to\infty} u(x,t) = 0$ implies $\lim_{x\to\infty} U(x,s) = 0$, so we must define $c_2 = 0$. Consequently

$$U(x,s) = c_1 e^{-(s/a)x} + \frac{v_0^2 F_0}{(a^2 - v_0^2)s^2} e^{-(s/v_0)x}.$$

The remaining boundary condition transforms into $U(0,s) = 0$. From this we find

$$c_1 = -v_0^2 F_0/(a^2 - v_0^2)s^2.$$

Therefore, by the second translation theorem

$$U(x,s) = -\frac{v_0^2 F_0}{(a^2 - v_0^2)s^2} e^{-(s/a)x} + \frac{v_0^2 F_0}{(a^2 - v_0^2)s^2} e^{-(s/v_0)x}$$

and

$$\begin{aligned}
u(x,t) &= \frac{v_0^2 F_0}{a^2 - v_0^2}\left[\mathscr{L}^{-1}\left\{\frac{e^{-(x/v_0)s}}{s^2}\right\} - \mathscr{L}^{-1}\left\{\frac{e^{-(x/a)s}}{s^2}\right\}\right] \\
&= \frac{v_0^2 F_0}{a^2 - v_0^2}\left[\left(t - \frac{x}{v_0}\right)\mathscr{U}\left(t - \frac{x}{v_0}\right) - \left(t - \frac{x}{a}\right)\mathscr{U}\left(t - \frac{x}{a}\right)\right].
\end{aligned}$$

(b) In the case when $v_0 = a$ the solution of the transformed equation is

$$U(x,s) = c_1 e^{-(s/a)x} + c_2 e^{(s/a)x} - \frac{F_0}{2as} x e^{-(s/a)x}.$$

The usual analysis then leads to $c_1 = 0$ and $c_2 = 0$. Therefore

$$U(x,s) = -\frac{F_0}{2as} x e^{-(s/a)x}$$

and

$$u(x,t) = -\frac{xF_0}{2a} \mathscr{L}^{-1}\left\{\frac{e^{-(x/a)s}}{s}\right\} = -\frac{xF_0}{2a} \mathscr{U}\left(t - \frac{x}{a}\right).$$

Exercises 14.3

3. From formulas (5) and (6) in the text,

$$\begin{aligned} A(\alpha) &= \int_0^3 x \cos \alpha x \, dx \\ &= \frac{x \sin \alpha x}{\alpha}\Big|_0^3 - \frac{1}{\alpha}\int_0^3 \sin \alpha x \, dx \\ &= \frac{3 \sin 3\alpha}{\alpha} + \frac{\cos \alpha x}{\alpha^2}\Big|_0^3 \\ &= \frac{3\alpha \sin 3\alpha + \cos 3\alpha - 1}{\alpha^2} \end{aligned}$$

and

$$\begin{aligned} B(\alpha) &= \int_0^3 x \sin \alpha x \, dx \\ &= -\frac{x \cos \alpha x}{\alpha}\Big|_0^3 + \frac{1}{\alpha}\int_0^3 \cos \alpha x \, dx \\ &= -\frac{3 \cos 3\alpha}{\alpha} + \frac{\sin \alpha x}{\alpha^2}\Big|_0^3 \\ &= \frac{\sin 3\alpha - 3\alpha \cos 3\alpha}{\alpha^2}. \end{aligned}$$

Hence

$$f(x)=\frac{1}{\pi}\int_0^\infty \frac{(3\alpha\sin 3\alpha+\cos 3\alpha-1)\cos\alpha x+(\sin 3\alpha-3\alpha\cos 3\alpha)\sin\alpha x}{\alpha^2}\,d\alpha$$

$$=\frac{1}{\pi}\int_0^\infty \frac{3\alpha(\sin 3\alpha\cos\alpha x-\cos 3\alpha\sin\alpha x)+\cos 3\alpha\cos\alpha x+\sin 3\alpha\sin\alpha x-\cos\alpha x}{\alpha^2}\,d\alpha$$

$$=\frac{1}{\pi}\int_0^\infty \frac{3\alpha\sin\alpha(3-x)+\cos\alpha(3-x)-\cos\alpha x}{\alpha^2}\,d\alpha.$$

6. From formulas (5) and (6) in the text,

$$A(\alpha)=\int_{-1}^1 e^x\cos\alpha x\,dx$$

$$=\frac{e(\cos\alpha+\alpha\sin\alpha)-e^{-1}(\cos\alpha-\alpha\sin\alpha)}{1+\alpha^2}$$

$$=\frac{2(\sinh 1)\cos\alpha-2\alpha(\cosh 1)\sin\alpha}{1+\alpha^2}$$

and

$$B(\alpha)=\int_{-1}^1 e^x\sin\alpha x\,dx$$

$$=\frac{e(\sin\alpha-\alpha\cos\alpha)-e^{-1}(-\sin\alpha-\alpha\cos\alpha)}{1+\alpha^2}$$

$$=\frac{2(\cosh 1)\sin\alpha-2\alpha(\sinh 1)\cos\alpha}{1+\alpha^2}.$$

Hence

$$f(x)=\frac{1}{\pi}\int_0^\infty [A(\alpha)\cos\alpha x+B(\alpha)\sin\alpha x]\,d\alpha.$$

9. The function is even. Thus from formula (9) in the text

$$A(\alpha)=\int_0^\pi x\cos\alpha x\,dx=\frac{x\sin\alpha x}{\alpha}\bigg|_0^\pi-\frac{1}{\alpha}\int_0^\pi\sin\alpha x\,dx$$

$$=\frac{\pi\alpha\sin\pi\alpha}{\alpha}+\frac{1}{\alpha^2}\cos\alpha x\bigg|_0^\pi=\frac{\pi\alpha\sin\pi\alpha+\cos\pi\alpha-1}{\alpha^2}.$$

Hence from formula (8) in the text

$$f(x)=\frac{2}{\pi}\int_0^\infty \frac{(\pi\alpha\sin\pi\alpha+\cos\pi\alpha-1)\cos\alpha x}{\alpha^2}\,d\alpha.$$

12. The function is odd. Thus from formula (11) in the text

$$B(\alpha)=\int_0^\infty xe^{-x}\sin\alpha x\,dx.$$

Now recall

$$\mathscr{L}\{t \sin kt\} = -\frac{d}{ds}\mathscr{L}\{\sin kt\} = 2ks/(s^2+k^2)^2.$$

If we set $s = 1$ and $k = \alpha$ we obtain

$$B(\alpha) = \frac{2\alpha}{(1+\alpha^2)^2}.$$

Hence from formula (10) in the text

$$f(x) = \frac{4}{\pi}\int_0^\infty \frac{\alpha \sin \alpha x}{(1+\alpha^2)^2}\,d\alpha.$$

15. For the cosine integral,

$$A(\alpha) = \int_0^\infty xe^{-2x}\cos \alpha x\,dx.$$

But we know

$$\mathscr{L}\{t \cos kt\} = -\frac{d}{ds}\frac{s}{(s^2+k^2)} = \frac{(s^2-k^2)}{(s^2+k^2)^2}.$$

If we set $s = 2$ and $k = \alpha$ we obtain

$$A(\alpha) = \frac{4-\alpha^2}{(4+\alpha^2)^2}.$$

Hence

$$f(x) = \frac{2}{\pi}\int_0^\infty \frac{(4-\alpha^2)\cos \alpha x}{(4+\alpha^2)^2}\,d\alpha.$$

For the sine integral,

$$B(\alpha) = \int_0^\infty xe^{-2x}\sin \alpha x\,dx.$$

From Problem 12, we know

$$\mathscr{L}\{t \sin kt\} = \frac{2ks}{(s^2+k^2)^2}.$$

If we set $s = 2$ and $k = \alpha$ we obtain

$$B(\alpha) = \frac{4\alpha}{(4+\alpha^2)^2}.$$

Hence

$$f(x) = \frac{8}{\pi}\int_0^\infty \frac{\alpha \sin \alpha x}{(4+\alpha^2)^2}\,d\alpha.$$

18. From the formula for sine integral of $f(x)$ we have

$$f(x) = \frac{2}{\pi}\int_0^\infty \left(\int_0^\infty f(x)\sin\alpha x\,dx\right)\sin\alpha x\,dx$$
$$= \frac{2}{\pi}\left[\int_0^1 1\cdot\sin\alpha x\,d\alpha + \int_1^\infty 0\cdot\sin\alpha x\,d\alpha\right]$$
$$= \frac{2}{\pi}\frac{(-\cos\alpha x)}{x}\bigg|_0^1$$
$$= \frac{2}{\pi}\frac{1-\cos x}{x}.$$

Exercises 14.4

For the boundary-value problems in this section it is sometimes useful to note that the identities

$$e^{i\alpha} = \cos\alpha + i\sin\alpha \quad\text{and}\quad e^{-i\alpha} = \cos\alpha - i\sin\alpha$$

imply

$$e^{i\alpha} + e^{-i\alpha} = 2\cos\alpha \quad\text{and}\quad e^{i\alpha} - e^{-i\alpha} = 2i\sin\alpha$$

3. Using the Fourier transform, the partial differential equation equation becomes

$$\frac{dU}{dt} + k\alpha^2 U = 0 \qquad\text{and so}\qquad U(\alpha,t) = ce^{-k\alpha^2 t}.$$

Now

$$\{u(x,0)\} = U(\alpha,0) = \sqrt{\pi}\,e^{-\alpha^2/4}$$

by the given result. This gives $c = \sqrt{\pi}\,e^{-\alpha^2/4}$ and so

$$U(\alpha,t) = \sqrt{\pi}\,e^{-(\frac{1}{4}+kt)\alpha^2}.$$

Using the given Fourier transform again we obtain

$$u(x,t) = \sqrt{\pi}\quad{}^{-1}\{e^{-(\frac{1+4kt}{4})\alpha^2}\} = \frac{1}{\sqrt{1+4kt}}e^{-x^2/(1+4kt)}.$$

6. The solution of Problem 5 can be written

$$u(x,t) = \frac{2u_0}{\pi}\int_0^\infty \frac{\sin\alpha x}{\alpha}\,d\alpha - \frac{2u_0}{\pi}\int_0^\infty \frac{\sin\alpha x}{\alpha}e^{-k\alpha^2 t}\,d\alpha.$$

Using $\int_0^\infty \frac{\sin\alpha x}{\alpha}\,d\alpha = \pi/2$ the last line becomes

$$u(x,t) = u_0 - \frac{2u_0}{\pi}\int_0^\infty \frac{\sin\alpha x}{\alpha}e^{-k\alpha^2 t}\,d\alpha.$$

9. Using the Fourier cosine transform we find

$$U(\alpha,t) = ce^{-k\alpha^2 t}.$$

Now

$$c\{u(x,0)\} = \int_0^1 \cos\alpha x\, dx = \frac{\sin\alpha}{\alpha} = U(\alpha,0).$$

From this we obtain $c = (\sin\alpha)/\alpha$ and so

$$U(\alpha,t) = \frac{\sin\alpha}{\alpha}\, e^{-k\alpha^2 t}$$

and

$$u(x,t) = \frac{2}{\pi}\int_0^\infty \frac{\sin\alpha}{\alpha}\, e^{-k\alpha^2 t}\cos\alpha x\, d\alpha.$$

12. Using the Fourier sine transform we obtain

$$U(\alpha,t) = c_1\cos\alpha at + c_2\sin\alpha at.$$

Now

$$s\{u(x,0)\} = \quad \{xe^{-x}\} = \int_0^\infty xe^{-x}\sin\alpha x\, dx = \frac{2\alpha}{(1+\alpha^2)^2} = U(\alpha,0).$$

Also,

$$s\{u_t(x,0)\} = \left.\frac{dU}{dt}\right|_{t=0} = 0.$$

This last condition gives $c_2 = 0$. Then $U(\alpha,0) = 2\alpha/(1+\alpha^2)^2$ yields $c_1 = 2\alpha/(1+\alpha^2)^2$. Therefore

$$U(\alpha,t) = \frac{2\alpha}{(1+\alpha^2)^2}\cos\alpha at$$

and

$$u(x,t) = \frac{4}{\pi}\int_0^\infty \frac{\alpha\cos\alpha at}{(1+\alpha^2)^2}\sin\alpha x\, d\alpha.$$

15. Using the Fourier cosine transform with respect to x gives

$$U(\alpha,y) = c_1e^{-\alpha y} + c_2e^{\alpha y}.$$

Since we expect $u(x,y)$ to be bounded as $y \to \infty$ we define $c_2 = 0$. Thus

$$U(\alpha,y) = c_1e^{-\alpha y}.$$

Now

$$c\{u(x,0)\} = \int_0^1 50\cos\alpha x\, dx = 50\,\frac{\sin\alpha}{\alpha}$$

and so

$$U(\alpha,y) = 50\,\frac{\sin\alpha}{\alpha}\, e^{-\alpha y}$$

and

$$u(x,y) = \frac{100}{\pi}\int_0^\infty \frac{\sin\alpha}{\alpha} e^{-\alpha y}\cos\alpha x\, d\alpha.$$

18. The domain of y and the boundary condition at $y = 0$ suggest that we use a Fourier cosine transform. The transformed equation is

$$\frac{d^2U}{dx^2} - \alpha^2 U - u_y(x,0) = 0 \quad \text{or} \quad \frac{d^2U}{dx^2} - \alpha^2 U = 0.$$

Because the domain of the variable x is a finite interval we choose to write the general solution of the latter equation as

$$U(x,\alpha) = c_1\cosh\alpha x + c_2\sinh\alpha x.$$

Now $U(0,\alpha) = F(\alpha)$, where $F(\alpha)$ is the Fourier cosine transform of $f(y)$, and $U'(\pi,\alpha) = 0$ imply $c_1 = F(\alpha)$ and $c_2 = -F(\alpha)\sinh\alpha\pi/\cosh\alpha\pi$. Thus

$$U(x,\alpha) = F(\alpha)\cosh\alpha x - F(\alpha)\frac{\sinh\alpha\pi}{\cosh\alpha\pi}\sinh\alpha x = F(\alpha)\frac{\cosh\alpha(\pi - x)}{\cosh\alpha\pi}.$$

Using the inverse transform we find that a solution to the problem is

$$u(x,y) = \frac{2}{\pi}\int_0^\infty F(\alpha)\frac{\cosh\alpha(\pi - x)}{\cosh\alpha\pi}\cos\alpha y\, d\alpha.$$

21. Using the Fourier transform with respect to x gives

$$U(\alpha,y) = c_1\cosh\alpha y + c_2\sinh\alpha y.$$

The transform of the boundary condition $\left.\frac{\partial u}{\partial y}\right|_{y=0} = 0$ is $\left.\frac{dU}{dy}\right|_{y=0} = 0$. This condition gives $c_2 = 0$. Hence

$$U(\alpha,y) = c_1\cosh\alpha y.$$

Now by the given information the transform of the boundary condition $u(x,1) = e^{-x^2}$ is $U(\alpha,1) = \sqrt{\pi}\,e^{-\alpha^2/4}$. This condition then gives $c_1 = \sqrt{\pi}\,e^{-\alpha^2/4}\cosh\alpha$. Therefore

$$U(\alpha,y) = \sqrt{\pi}\,\frac{e^{-\alpha^2/4}\cosh\alpha y}{\cosh\alpha}$$

and

$$U(x,y) = \frac{1}{2\sqrt{\pi}}\int_{-\infty}^\infty \frac{e^{-\alpha^2/4}\cosh\alpha y}{\cosh\alpha} e^{-i\alpha x} d\alpha$$

$$= \frac{1}{2\sqrt{\pi}}\int_{-\infty}^\infty \frac{e^{-\alpha^2/4}\cosh\alpha y}{\cosh\alpha}\cos\alpha x\, d\alpha$$

$$= \frac{1}{\sqrt{\pi}}\int_0^\infty \frac{e^{-\alpha^2/4}\cosh\alpha y}{\cosh\alpha}\cos\alpha x\, d\alpha.$$

24. Using integration by parts,

$$\mathcal{F}_c\{f'(x)\} = \int_0^\infty f'(x)\cos\alpha x\,dx = f(x)\cos\alpha x\,\Big|_0^\infty + \alpha\int_0^\infty f(x)\sin\alpha x\,dx.$$

If we assume that $f(x)\cos\alpha x \to 0$ as $x\to\infty$ and that f is bounded at $x=0$, we obtain

$$\mathcal{F}_c\{f'(x)\} = -f(0) + \alpha\,\mathcal{F}_s\{f(x)\}.$$

Chapter 14 Review Exercises

3. The Laplace transform gives

$$U(x,s) = c_1 e^{-\sqrt{s+h}\,x} + c_2 e^{\sqrt{s+h}\,x} + \frac{u_0}{s+h}.$$

The condition $\lim_{x\to\infty}\partial u/\partial x = 0$ implies $\lim_{x\to\infty} dU/dx = 0$ and so we define $c_2 = 0$. Thus

$$U(x,s) = c_1 e^{-\sqrt{s+h}\,x} + \frac{u_0}{s+h}.$$

The condition $U(0,s)=0$ then gives $c_1 = -u_0/(s+h)$ and so

$$U(x,s) = \frac{u_0}{s+h} - u_0\,\frac{e^{-\sqrt{s+h}\,x}}{s+h}.$$

With the help of the first translation theorem we then obtain

$$u(x,t) = u_0\,\mathscr{L}^{-1}\left\{\frac{1}{s+h}\right\} - u_0\,\mathscr{L}^{-1}\left\{\frac{e^{-\sqrt{s+h}\,x}}{s+h}\right\} = u_0e^{-ht} - u_0e^{-ht}\operatorname{erfc}\left(\frac{x}{2\sqrt{t}}\right)$$

$$= u_0e^{-ht}\left[1-\operatorname{erfc}\left(\frac{x}{2\sqrt{t}}\right)\right] = u_0e^{-ht}\operatorname{erf}\left(\frac{x}{2\sqrt{t}}\right).$$

6. The Laplace transform and undetermined coefficients gives

$$U(x,s) = c_1\cosh sx + c_2\sinh sx + \frac{s-1}{s^2+\pi^2}\sin\pi x.$$

The conditions $U(0,s)=0$ and $U(1,s)=0$ give, in turn, $c_1=0$ and $c_2=0$. Thus

$$U(x,s) = \frac{s-1}{s^2+\pi^2}\sin\pi x$$

and

$$u(x,t) = \sin\pi x\,\mathscr{L}^{-1}\left\{\frac{s}{s^2+\pi^2}\right\} - \frac{1}{\pi}\sin\pi x\,\mathscr{L}^{-1}\left\{\frac{\pi}{s^2+\pi^2}\right\}$$

$$= (\sin\pi x)\cos\pi t - \frac{1}{\pi}(\sin\pi x)\sin\pi t.$$

9. We solve the two problems

$$\frac{\partial^2 u_1}{\partial x^2} + \frac{\partial^2 u_1}{\partial y^2} = 0, \quad x > 0, \quad y > 0,$$

$$u_1(0,y) = 0, \quad y > 0,$$

$$u_1(x,0) = \begin{cases} 100, & 0 < x < 1 \\ 0, & x > 1 \end{cases}$$

and

$$\frac{\partial^2 u_2}{\partial x^2} + \frac{\partial^2 u_2}{\partial y^2} = 0, \quad x > 0, \quad y > 0,$$

$$u_2(0,y) = \begin{cases} 50, & 0 < y < 1 \\ 0, & y > 1 \end{cases}$$

$$u_2(x,0) = 0.$$

Using the Fourier sine transform with respect to x we find

$$u_1(x,y) = \frac{200}{\pi} \int_0^\infty \left(\frac{1 - \cos\alpha}{\alpha} \right) e^{-\alpha y} \sin \alpha x \, d\alpha.$$

Using the Fourier sine transform with respect to y we find

$$u_2(x,y) = \frac{100}{\pi} \int_0^\infty \left(\frac{1 - \cos\alpha}{\alpha} \right) e^{-\alpha x} \sin \alpha y \, d\alpha.$$

The solution of the problem is then

$$u(x,y) = u_1(x,y) + u_2(x,y).$$

12. Using the Laplace transform gives

$$U(x,s) = c_1 \cosh \sqrt{s}\, x + c_2 \sinh \sqrt{s}\, x.$$

The condition $u(0,t) = u_0$ transforms into $U(0,s) = u_0/s$. This gives $c_1 = u_0/s$. The condition $u(1,t) = u_0$ transforms into $U(1,s) = u_0/s$. This implies that $c_2 = u_0(1 - \cosh\sqrt{s})/s \sinh\sqrt{s}$. Hence

$$\begin{aligned} U(x,s) &= \frac{u_0}{s} \cosh \sqrt{s}\, x + u_0 \left[\frac{1 - \cosh\sqrt{s}}{s \sinh\sqrt{s}} \right] \sinh \sqrt{s}\, x \\ &= u_0 \left[\frac{\sinh\sqrt{s} \cosh\sqrt{s}\, x - \cosh \sinh\sqrt{s} \sinh\sqrt{s}\, x + \sinh\sqrt{s}\, x}{s \sinh\sqrt{s}} \right] \\ &= u_0 \left[\frac{\sinh\sqrt{s}\,(1-x) + \sinh\sqrt{s}\, x}{s \sinh\sqrt{s}} \right] \\ &= u_0 \left[\frac{\sinh\sqrt{s}\,(1-x)}{s \sinh\sqrt{s}} + \frac{\sinh\sqrt{s}\, x}{s \sinh\sqrt{s}} \right] \end{aligned}$$

and

$$\begin{aligned}
u(x,t) &= u_0\left[\mathscr{L}^{-1}\left\{\frac{\sinh\sqrt{s}\,(1-x)}{s\sinh\sqrt{s}}\right\}+\mathscr{L}^{-1}\left\{\frac{\sinh\sqrt{s}\,x}{s\sinh\sqrt{s}}\right\}\right]\\
&= u_0\sum_{n=0}^{\infty}\left[\operatorname{erf}\left(\frac{2n+2-x}{2\sqrt{t}}\right)-\operatorname{erf}\left(\frac{2n+x}{2\sqrt{t}}\right)\right]\\
&\quad+u_0\sum_{n=0}^{\infty}\left[\operatorname{erf}\left(\frac{2n+1+x}{2\sqrt{t}}\right)-\operatorname{erf}\left(\frac{2n+1-x}{2\sqrt{t}}\right)\right].
\end{aligned}$$

15 Numerical Methods for Partial Differential Equations

Exercises 15.1

3. The figure shows the values of $u(x, y)$ along the boundary. We need to determine u_{11}, u_{21}, u_{12}, and u_{22}. By symmetry $u_{11} = u_{21}$ and $u_{12} = u_{22}$. The system is

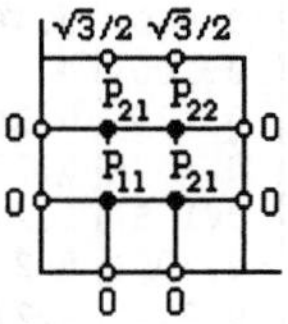

$$u_{21} + u_{12} + 0 + 0 - 4u_{11} = 0$$

$$0 + u_{22} + u_{11} + 0 - 4u_{21} = 0 \qquad\qquad 3u_{11} + u_{12} = 0$$

$$\text{or}$$

$$u_{22} + \sqrt{3}/2 + 0 + u_{11} - 4u_{12} = 0 \qquad\qquad u_{11} - 3u_{12} = -\frac{\sqrt{3}}{2}.$$

$$0 + \sqrt{3}/2 + u_{12} + u_{21} - 4u_{22} = 0$$

Solving we obtain $u_{11} = u_{21} = \sqrt{3}/16$ and $u_{12} = u_{22} = 3\sqrt{3}/16$.

6. For Gauss-Seidel the coefficients of the unknowns u_{11}, u_{21}, u_{31}, u_{12}, u_{22}, u_{32}, u_{13}, u_{23}, u_{33} are shown in the matrix

$$\begin{bmatrix} 0 & .25 & 0 & .25 & 0 & 0 & 0 & 0 & 0 \\ .25 & 0 & .25 & 0 & .25 & 0 & 0 & 0 & 0 \\ 0 & .25 & 0 & 0 & 0 & .25 & 0 & 0 & 0 \\ .25 & 0 & 0 & 0 & .25 & 0 & .25 & 0 & 0 \\ 0 & .25 & 0 & .25 & 0 & .25 & 0 & .25 & 0 \\ 0 & 0 & .25 & 0 & .25 & 0 & 0 & 0 & .25 \\ 0 & 0 & 0 & .25 & 0 & 0 & 0 & .25 & 0 \\ 0 & 0 & 0 & 0 & .25 & 0 & .25 & 0 & .25 \\ 0 & 0 & 0 & 0 & 0 & .25 & 0 & .25 & 0 \end{bmatrix}.$$

The constant terms are 7.5, 5, 20, 10, 0, 15, 17.5, 5, 27.5. We use 32.5 as the initial guess for each variable. Then $u_{11} = 21.92$, $u_{21} = 28.30$, $u_{31} = 38.17$, $u_{12} = 29.38$, $u_{22} = 33.13$, $u_{32} = 44.38$, $u_{13} = 22.46$, $u_{23} = 30.45$, and $u_{33} = 46.21$.

9. Identifying $u_{ij} = u(x, t)$ the difference equation is given by

$$\frac{1}{h^2}(u_{i+1,j} - 2u_{ij} + u_{i-1,j}) = \frac{1}{k}(u_{i,j+1} - u_{ij})$$

or

$$u_{i,j+1} = \left(1 - \frac{2k}{h^2}\right)u_{ij} + \frac{k}{h^2}(u_{i+1,j} + u_{i-1,j}).$$

Exercises 15.2

3. We identify $c = 1$, $a = 2$, $T = 1$, $n = 8$, and $m = 40$. Then $h = 2/8 = 0.25$, $k = 1/40 = 0.025$, and $\lambda = 2/5 = 0.4$.

TIME	X=0.25	X=0.50	X=0.75	X=1.00	X=1.25	X=1.50	X=1.75
0.000	1.0000	1.0000	1.0000	1.0000	0.0000	0.0000	0.0000
0.025	0.7074	0.9520	0.9566	0.7444	0.2545	0.0371	0.0053
0.050	0.5606	0.8499	0.8685	0.6633	0.3303	0.1034	0.0223
0.075	0.4684	0.7473	0.7836	0.6191	0.3614	0.1529	0.0462
0.100	0.4015	0.6577	0.7084	0.5837	0.3753	0.1871	0.0684
0.125	0.3492	0.5821	0.6428	0.5510	0.3797	0.2101	0.0861
0.150	0.3069	0.5187	0.5857	0.5199	0.3778	0.2247	0.0990
0.175	0.2721	0.4652	0.5359	0.4901	0.3716	0.2329	0.1078
0.200	0.2430	0.4198	0.4921	0.4617	0.3622	0.2362	0.1132
0.225	0.2186	0.3809	0.4533	0.4348	0.3507	0.2358	0.1160
0.250	0.1977	0.3473	0.4189	0.4093	0.3378	0.2327	0.1166
0.275	0.1798	0.3181	0.3881	0.3853	0.3240	0.2275	0.1157
0.300	0.1643	0.2924	0.3604	0.3626	0.3097	0.2208	0.1136
0.325	0.1507	0.2697	0.3353	0.3412	0.2953	0.2131	0.1107
0.350	0.1387	0.2495	0.3125	0.3211	0.2808	0.2047	0.1071
0.375	0.1281	0.2313	0.2916	0.3021	0.2666	0.1960	0.1032
0.400	0.1187	0.2150	0.2725	0.2843	0.2528	0.1871	0.0989
0.425	0.1102	0.2002	0.2549	0.2675	0.2393	0.1781	0.0946
0.450	0.1025	0.1867	0.2387	0.2517	0.2263	0.1692	0.0902
0.475	0.0955	0.1743	0.2236	0.2368	0.2139	0.1606	0.0858
0.500	0.0891	0.1630	0.2097	0.2228	0.2020	0.1521	0.0814
0.525	0.0833	0.1525	0.1967	0.2096	0.1906	0.1439	0.0772
0.550	0.0779	0.1429	0.1846	0.1973	0.1798	0.1361	0.0731
0.575	0.0729	0.1339	0.1734	0.1856	0.1696	0.1285	0.0691
0.600	0.0683	0.1256	0.1628	0.1746	0.1598	0.1214	0.0653
0.625	0.0641	0.1179	0.1530	0.1643	0.1506	0.1145	0.0617
0.650	0.0601	0.1106	0.1438	0.1546	0.1419	0.1080	0.0582
0.675	0.0564	0.1039	0.1351	0.1455	0.1336	0.1018	0.0549
0.700	0.0530	0.0976	0.1270	0.1369	0.1259	0.0959	0.0518
0.725	0.0497	0.0917	0.1194	0.1288	0.1185	0.0904	0.0488
0.750	0.0467	0.0862	0.1123	0.1212	0.1116	0.0852	0.0460
0.775	0.0439	0.0810	0.1056	0.1140	0.1050	0.0802	0.0433
0.800	0.0413	0.0762	0.0993	0.1073	0.0989	0.0755	0.0408
0.825	0.0388	0.0716	0.0934	0.1009	0.0931	0.0711	0.0384
0.850	0.0365	0.0674	0.0879	0.0950	0.0876	0.0669	0.0362
0.875	0.0343	0.0633	0.0827	0.0894	0.0824	0.0630	0.0341
0.900	0.0323	0.0596	0.0778	0.0841	0.0776	0.0593	0.0321
0.925	0.0303	0.0560	0.0732	0.0791	0.0730	0.0558	0.0302
0.950	0.0285	0.0527	0.0688	0.0744	0.0687	0.0526	0.0284
0.975	0.0268	0.0496	0.0647	0.0700	0.0647	0.0495	0.0268
1.000	0.0253	0.0466	0.0609	0.0659	0.0608	0.0465	0.0252

(x,y)	exact	approx	abs error
(0.25,0.1)	0.3794	0.4015	0.0221
(1,0.5)	0.1854	0.2228	0.0374
(1.5,0.8)	0.0623	0.0755	0.0132

6. **(a)** We identify $c = 15/88 \approx 0.1705$, $a = 20$, $T = 10$, $n = 10$, and $m = 10$. Then $h = 2$, $k = 1$, and $\lambda = 15/352 \approx 0.0426$.

TIME	X=2	X=4	X=6	X=8	X=10	X=12	X=14	X=16	X=18
0	30.0000	30.0000	30.0000	30.0000	30.0000	30.0000	30.0000	30.0000	30.0000
1	28.7216	30.0000	30.0000	30.0000	30.0000	30.0000	30.0000	30.0000	28.7216
2	27.5521	29.9455	30.0000	30.0000	30.0000	30.0000	30.0000	29.9455	27.5521
3	26.4800	29.8459	29.9977	30.0000	30.0000	30.0000	29.9977	29.8459	26.4800
4	25.4951	29.7089	29.9913	29.9999	30.0000	29.9999	29.9913	29.7089	25.4951
5	24.5882	29.5414	29.9796	29.9995	30.0000	29.9995	29.9796	29.5414	24.5882
6	23.7515	29.3490	29.9618	29.9987	30.0000	29.9987	29.9618	29.3490	23.7515
7	22.9779	29.1365	29.9373	29.9972	29.9998	29.9972	29.9373	29.1365	22.9779
8	22.2611	28.9082	29.9057	29.9948	29.9996	29.9948	29.9057	28.9082	22.2611
9	21.5958	28.6675	29.8670	29.9912	29.9992	29.9912	29.8670	28.6675	21.5958
10	20.9768	28.4172	29.8212	29.9862	29.9985	29.9862	29.8212	28.4172	20.9768

(b) We identify $c = 15/88 \approx 0.1705$, $a = 50$, $T = 10$, $n = 10$, and $m = 10$. Then $h = 5$, $k = 1$, and $\lambda = 3/440 \approx 0.0068$.

TIME	X=5	X=10	X=15	X=20	X=25	X=30	X=35	X=40	X=45
0	30.0000	30.0000	30.0000	30.0000	30.0000	30.0000	30.0000	30.0000	30.0000
1	29.7955	30.0000	30.0000	30.0000	30.0000	30.0000	30.0000	30.0000	29.7955
2	29.5937	29.9986	30.0000	30.0000	30.0000	30.0000	30.0000	29.9986	29.5937
3	29.3947	29.9959	30.0000	30.0000	30.0000	30.0000	30.0000	29.9959	29.3947
4	29.1984	29.9918	30.0000	30.0000	30.0000	30.0000	30.0000	29.9918	29.1984
5	29.0047	29.9864	29.9999	30.0000	30.0000	30.0000	29.9999	29.9864	29.0047
6	28.8136	29.9798	29.9998	30.0000	30.0000	30.0000	29.9998	29.9798	28.8136
7	28.6251	29.9720	29.9997	30.0000	30.0000	30.0000	29.9997	29.9720	28.6251
8	28.4391	29.9630	29.9995	30.0000	30.0000	30.0000	29.9995	29.9630	28.4391
9	28.2556	29.9529	29.9992	30.0000	30.0000	30.0000	29.9992	29.9529	28.2556
10	28.0745	29.9416	29.9989	30.0000	30.0000	30.0000	29.9989	29.9416	28.0745

(c) We identify $c = 50/27 \approx 1.8519$, $a = 20$, $T = 10$, $n = 10$, and $m = 10$. Then $h = 2$, $k = 1$, and $\lambda = 25/54 \approx 0.4630$.

TIME	X=2	X=4	X=6	X=8	X=10	X=12	X=14	X=16	X=18
0	18.0000	32.0000	42.0000	48.0000	50.0000	48.0000	42.0000	32.0000	18.0000
1	16.1481	30.1481	40.1481	46.1481	48.1481	46.1481	40.1481	30.1481	16.1481
2	15.1536	28.2963	38.2963	44.2963	46.2963	44.2963	38.2963	28.2963	15.1536
3	14.2226	26.8414	36.4444	42.4444	44.4444	42.4444	36.4444	26.8414	14.2226
4	13.4801	25.4452	34.7764	40.5926	42.5926	40.5926	34.7764	25.4452	13.4801
5	12.7787	24.2258	33.1491	38.8258	40.7407	38.8258	33.1491	24.2258	12.7787
6	12.1622	23.0574	31.6460	37.0842	38.9677	37.0842	31.6460	23.0574	12.1622
7	11.5756	21.9895	30.1875	35.4385	37.2238	35.4385	30.1875	21.9895	11.5756
8	11.0378	20.9636	28.8232	33.8340	35.5707	33.8340	28.8232	20.9636	11.0378
9	10.5230	20.0070	27.5043	32.3182	33.9626	32.3182	27.5043	20.0070	10.5230
10	10.0420	19.0872	26.2620	30.8509	32.4400	30.8509	26.2620	19.0872	10.0420

(d) We identify $c = 260/159 \approx 1.6352$, $a = 100$, $T = 10$, $n = 10$, and $m = 10$. Then $h = 10$, $k = 1$, and $\lambda = 13/795 \approx 00164$.

TIME	X=10	X=20	X=30	X=40	X=50	X=60	X=70	X=80	X=90
0	8.0000	16.0000	24.0000	32.0000	40.0000	32.0000	24.0000	16.0000	8.0000
1	8.0000	16.0000	24.0000	32.0000	39.7384	32.0000	24.0000	16.0000	8.0000
2	8.0000	16.0000	24.0000	31.9957	39.4853	31.9957	24.0000	16.0000	8.0000
3	8.0000	16.0000	23.9999	31.9874	39.2403	31.9874	23.9999	16.0000	8.0000
4	8.0000	16.0000	23.9997	31.9754	39.0031	31.9754	23.9997	16.0000	8.0000
5	8.0000	16.0000	23.9993	31.9599	38.7733	31.9599	23.9993	16.0000	8.0000
6	8.0000	16.0000	23.9987	31.9412	38.5505	31.9412	23.9987	16.0000	8.0000
7	8.0000	16.0000	23.9978	31.9194	38.3343	31.9194	23.9978	16.0000	8.0000
8	8.0000	15.9999	23.9965	31.8947	38.1245	31.8947	23.9965	15.9999	8.0000
9	8.0000	15.9999	23.9949	31.8675	37.9208	31.8675	23.9949	15.9999	8.0000
10	8.0000	15.9998	23.9929	31.8377	37.7228	31.8377	23.9929	15.9998	8.0000

9. (a) We identify $c = 15/88 \approx 0.1705$, $a = 20$, $T = 10$, $n = 10$, and $m = 10$. Then $h = 2$, $k = 1$, and $\lambda = 15/352 \approx 0.0426$.

TIME	X=2.00	X=4.00	X=6.00	X=8.00	X=10.00	X=12.00	X=14.00	X=16.00	X=18.00
0.00	30.0000	30.0000	30.0000	30.0000	30.0000	30.0000	30.0000	30.0000	30.0000
1.00	28.7733	29.9749	29.9995	30.0000	30.0000	30.0000	29.9998	29.9916	29.5911
2.00	27.6450	29.9037	29.9970	29.9999	30.0000	30.0000	29.9990	29.9679	29.2150
3.00	26.6051	29.7938	29.9911	29.9997	30.0000	29.9999	29.9970	29.9313	28.8684
4.00	25.6452	29.6517	29.9805	29.9991	30.0000	29.9997	29.9935	29.8839	28.5484
5.00	24.7573	29.4829	29.9643	29.9981	29.9999	29.9994	29.9881	29.8276	28.2524
6.00	23.9347	29.2922	29.9421	29.9963	29.9997	29.9988	29.9807	29.7641	27.9782
7.00	23.1711	29.0836	29.9134	29.9936	29.9995	29.9979	29.9711	29.6945	27.7237
8.00	22.4612	28.8606	29.8782	29.9899	29.9991	29.9966	29.9594	29.6202	27.4870
9.00	21.7999	28.6263	29.8362	29.9848	29.9985	29.9949	29.9454	29.5421	27.2666
10.00	21.1829	28.3831	29.7878	29.9783	29.9976	29.9927	29.9293	29.4610	27.0610

(b) We identify $c = 15/88 \approx 0.1705$, $a = 50$, $T = 10$, $n = 10$, and $m = 10$. Then $h = 5$, $k = 1$, and $\lambda = 3/440 \approx 0.0068$.

TIME	X=5.00	X=10.00	X=15.00	X=20.00	X=25.00	X=30.00	X=35.00	X=40.00	X=45.00
0.00	30.0000	30.0000	30.0000	30.0000	30.0000	30.0000	30.0000	30.0000	30.0000
1.00	29.7968	29.9993	30.0000	30.0000	30.0000	30.0000	30.0000	29.9998	29.9323
2.00	29.5964	29.9973	30.0000	30.0000	30.0000	30.0000	30.0000	29.9991	29.8655
3.00	29.3987	29.9939	30.0000	30.0000	30.0000	30.0000	30.0000	29.9980	29.7996
4.00	29.2036	29.9893	29.9999	30.0000	30.0000	30.0000	30.0000	29.9964	29.7345
5.00	29.0112	29.9834	29.9998	30.0000	30.0000	30.0000	29.9999	29.9945	29.6704
6.00	28.8212	29.9762	29.9997	30.0000	30.0000	30.0000	29.9999	29.9921	29.6071
7.00	28.6339	29.9679	29.9995	30.0000	30.0000	30.0000	29.9998	29.9893	29.5446
8.00	28.4490	29.9585	29.9992	30.0000	30.0000	30.0000	29.9997	29.9862	29.4830
9.00	28.2665	29.9479	29.9989	30.0000	30.0000	30.0000	29.9996	29.9827	29.4222
10.00	28.0864	29.9363	29.9986	30.0000	30.0000	30.0000	29.9995	29.9788	29.3621

(c) We identify $c = 50/27 \approx 1.8519$, $a = 20$, $T = 10$, $n = 10$, and $m = 10$. Then $h = 2$, $k = 1$, and $\lambda = 25/54 \approx 0.4630$.

TIME	X=2.00	X=4.00	X=6.00	X=8.00	X=10.00	X=12.00	X=14.00	X=16.00	X=18.00
0.00	18.0000	32.0000	42.0000	48.0000	50.0000	48.0000	42.0000	32.0000	18.0000
1.00	16.4489	30.1970	40.1562	46.1502	48.1531	46.1773	40.3274	31.2520	22.9449
2.00	15.3312	28.5350	38.3477	44.3130	46.3327	44.4671	39.0872	31.5755	24.6930
3.00	14.4219	27.0429	36.6090	42.5113	44.5759	42.9362	38.1976	31.7478	25.4131
4.00	13.6381	25.6913	34.9606	40.7728	42.9127	41.5716	37.4340	31.7086	25.6986
5.00	12.9409	24.4545	33.4091	39.1182	41.3519	40.3240	36.7033	31.5136	25.7663
6.00	12.3088	23.3146	31.9546	37.5566	39.8880	39.1565	35.9745	31.2134	25.7128
7.00	11.7294	22.2589	30.5939	36.0884	38.5109	38.0470	35.2407	30.8434	25.5871
8.00	11.1946	21.2785	29.3217	34.7092	37.2109	36.9834	34.5032	30.4279	25.4167
9.00	10.6987	20.3660	28.1318	33.4130	35.9801	35.9591	33.7660	29.9836	25.2181
10.00	10.2377	19.5150	27.0178	32.1929	34.8117	34.9710	33.0338	29.5224	25.0019

(d) We identify $c = 260/159 \approx 1.6352$, $a = 100$, $T = 10$, $n = 10$, and $m = 10$. Then $h = 10$, $k = 1$, and $\lambda = 13/795 \approx 00164$.

TIME	X=10.00	X=20.00	X=30.00	X=40.00	X=50.00	X=60.00	X=70.00	X=80.00	X=90.00
0.00	8.0000	16.0000	24.0000	32.0000	40.0000	32.0000	24.0000	16.0000	8.0000
1.00	8.0000	16.0000	24.0000	31.9979	39.7425	31.9979	24.0000	16.0026	8.3218
2.00	8.0000	16.0000	23.9999	31.9918	39.4932	31.9918	24.0000	16.0102	8.6333
3.00	8.0000	16.0000	23.9997	31.9820	39.2517	31.9820	24.0001	16.0225	8.9350
4.00	8.0000	16.0000	23.9993	31.9686	39.0175	31.9687	24.0002	16.0391	9.2272
5.00	8.0000	16.0000	23.9987	31.9520	38.7905	31.9520	24.0003	16.0599	9.5103
6.00	8.0000	15.9999	23.9978	31.9323	38.5701	31.9324	24.0005	16.0845	9.7846
7.00	8.0000	15.9999	23.9966	31.9097	38.3561	31.9098	24.0008	16.1126	10.0506
8.00	8.0000	15.9998	23.9950	31.8844	38.1483	31.8846	24.0012	16.1441	10.3084
9.00	8.0000	15.9997	23.9931	31.8566	37.9463	31.8569	24.0017	16.1786	10.5585
10.00	8.0000	15.9996	23.9908	31.8265	37.7499	31.8269	24.0023	16.2160	10.8012

Exercises 15.3

3. (a) Identifying $h = 1/5$ and $k = 0.5/10 = 0.05$ we see that $\lambda = 0.25$.

TIME	X=0.2	X=0.4	X=0.6	X=0.8
0.00	0.5878	0.9511	0.9511	0.5878
0.05	0.5808	0.9397	0.9397	0.5808
0.10	0.5599	0.9059	0.9059	0.5599
0.15	0.5256	0.8505	0.8505	0.5256
0.20	0.4788	0.7748	0.7748	0.4788
0.25	0.4206	0.6806	0.6806	0.4206
0.30	0.3524	0.5701	0.5701	0.3524
0.35	0.2757	0.4460	0.4460	0.2757
0.40	0.1924	0.3113	0.3113	0.1924
0.45	0.1046	0.1692	0.1692	0.1046
0.50	0.0142	0.0230	0.0230	0.0142

(b) Identifying $h = 1/5$ and $k = 0.5/20 = 0.025$ we see that $\lambda = 0.125$.

TIME	X=0.2	X=0.4	X=0.6	X=0.8
0.00	0.5878	0.9511	0.9511	0.5878
0.03	0.5860	0.9482	0.9482	0.5860
0.05	0.5808	0.9397	0.9397	0.5808
0.08	0.5721	0.9256	0.9256	0.5721
0.10	0.5599	0.9060	0.9060	0.5599
0.13	0.5445	0.8809	0.8809	0.5445
0.15	0.5257	0.8507	0.8507	0.5257
0.18	0.5039	0.8153	0.8153	0.5039
0.20	0.4790	0.7750	0.7750	0.4790
0.23	0.4513	0.7302	0.7302	0.4513
0.25	0.4209	0.6810	0.6810	0.4209
0.28	0.3879	0.6277	0.6277	0.3879
0.30	0.3527	0.5706	0.5706	0.3527
0.33	0.3153	0.5102	0.5102	0.3153
0.35	0.2761	0.4467	0.4467	0.2761
0.38	0.2352	0.3806	0.3806	0.2352
0.40	0.1929	0.3122	0.3122	0.1929
0.43	0.1495	0.2419	0.2419	0.1495
0.45	0.1052	0.1701	0.1701	0.1052
0.48	0.0602	0.0974	0.0974	0.0602
0.50	0.0149	0.0241	0.0241	0.0149

6. We identify $c = 24944.4$, $k = 0.00010022$ seconds $= 0.10022$ milliseconds, and $\lambda = 0.25$. Time in the table is expressed in milliseconds.

TIME	X=10	X=20	X=30	X=40	X=50
0.00000	0.2000	0.2667	0.2000	0.1333	0.0667
0.10022	0.1958	0.2625	0.2000	0.1333	0.0667
0.20045	0.1836	0.2503	0.1997	0.1333	0.0667
0.30067	0.1640	0.2307	0.1985	0.1333	0.0667
0.40089	0.1384	0.2050	0.1952	0.1332	0.0667
0.50111	0.1083	0.1744	0.1886	0.1328	0.0667
0.60134	0.0755	0.1407	0.1777	0.1318	0.0666
0.70156	0.0421	0.1052	0.1615	0.1295	0.0665
0.80178	0.0100	0.0692	0.1399	0.1253	0.0661
0.90201	-0.0190	0.0340	0.1129	0.1184	0.0654
1.00223	-0.0435	0.0004	0.0813	0.1077	0.0638
1.10245	-0.0626	-0.0309	0.0464	0.0927	0.0610
1.20268	-0.0758	-0.0593	0.0095	0.0728	0.0564
1.30290	-0.0832	-0.0845	-0.0278	0.0479	0.0493
1.40312	-0.0855	-0.1060	-0.0639	0.0184	0.0390
1.50334	-0.0837	-0.1237	-0.0974	-0.0150	0.0250
1.60357	-0.0792	-0.1371	-0.1275	-0.0511	0.0069
1.70379	-0.0734	-0.1464	-0.1533	-0.0882	-0.0152
1.80401	-0.0675	-0.1515	-0.1747	-0.1249	-0.0410
1.90424	-0.0627	-0.1528	-0.1915	-0.1595	-0.0694
2.00446	-0.0596	-0.1509	-0.2039	-0.1904	-0.0991
2.10468	-0.0585	-0.1467	-0.2122	-0.2165	-0.1283
2.20491	-0.0592	-0.1410	-0.2166	-0.2368	-0.1551
2.30513	-0.0614	-0.1349	-0.2175	-0.2507	-0.1772
2.40535	-0.0643	-0.1294	-0.2154	-0.2579	-0.1929
2.50557	-0.0672	-0.1251	-0.2105	-0.2585	-0.2005
2.60580	-0.0696	-0.1227	-0.2033	-0.2524	-0.1993
2.70602	-0.0709	-0.1219	-0.1942	-0.2399	-0.1889
2.80624	-0.0710	-0.1225	-0.1833	-0.2214	-0.1699
2.90647	-0.0699	-0.1236	-0.1711	-0.1972	-0.1435
3.00669	-0.0678	-0.1244	-0.1575	-0.1681	-0.1115
3.10691	-0.0649	-0.1237	-0.1425	-0.1348	-0.0761
3.20713	-0.0617	-0.1205	-0.1258	-0.0983	-0.0395
3.30736	-0.0583	-0.1139	-0.1071	-0.0598	-0.0042
3.40758	-0.0547	-0.1035	-0.0859	-0.0209	0.0279
3.50780	-0.0508	-0.0889	-0.0617	0.0171	0.0552
3.60803	-0.0460	-0.0702	-0.0343	0.0525	0.0767
3.70825	-0.0399	-0.0478	-0.0037	0.0840	0.0919
3.80847	-0.0318	-0.0221	0.0297	0.1106	0.1008
3.90870	-0.0211	0.0062	0.0648	0.1314	0.1041
4.00892	-0.0074	0.0365	0.1005	0.1464	0.1025
4.10914	0.0095	0.0680	0.1350	0.1558	0.0973
4.20936	0.0295	0.1000	0.1666	0.1602	0.0897
4.30959	0.0521	0.1318	0.1937	0.1606	0.0808
4.40981	0.0764	0.1625	0.2148	0.1581	0.0719
4.51003	0.1013	0.1911	0.2291	0.1538	0.0639
4.61026	0.1254	0.2164	0.2364	0.1485	0.0575
4.71048	0.1475	0.2373	0.2369	0.1431	0.0532
4.81070	0.1659	0.2526	0.2315	0.1379	0.0512
4.91093	0.1794	0.2611	0.2217	0.1331	0.0514
5.01115	0.1867	0.2620	0.2087	0.1288	0.0535

Chapter 15 Review Exercises

3. (a)

TIME	X=0.0	X=0.2	X=0.4	X=0.6	X=0.8	X=1.0
0.00	0.0000	0.2000	0.4000	0.6000	0.8000	0.0000
0.01	0.0000	0.2000	0.4000	0.6000	0.5500	0.0000
0.02	0.0000	0.2000	0.4000	0.5375	0.4250	0.0000
0.03	0.0000	0.2000	0.3844	0.4750	0.3469	0.0000
0.04	0.0000	0.1961	0.3609	0.4203	0.2922	0.0000
0.05	0.0000	0.1883	0.3346	0.3734	0.2512	0.0000

(b)

TIME	X=0.0	X=0.2	X=0.4	X=0.6	X=0.8	X=1.0
0.00	0.0000	0.2000	0.4000	0.6000	0.8000	0.0000
0.01	0.0000	0.2000	0.4000	0.6000	0.8000	0.0000
0.02	0.0000	0.2000	0.4000	0.6000	0.5500	0.0000
0.03	0.0000	0.2000	0.4000	0.5375	0.4250	0.0000
0.04	0.0000	0.2000	0.3844	0.4750	0.3469	0.0000
0.05	0.0000	0.1961	0.3609	0.4203	0.2922	0.0000

(c) The table in part (b) is the same as the table in part (a) shifted downward one row.

Appendix

Appendix I

3. If $t = x^3$, then $dt = 3x^2\,dx$ and $x^4\,dx = \frac{1}{3}t^{2/3}\,dt$. Now

$$\int_0^\infty x^4 e^{-x^3}\,dx = \int_0^\infty \frac{1}{3}t^{2/3}e^{-t}\,dt = \frac{1}{3}\int_0^\infty t^{2/3}e^{-t}\,dt$$
$$= \frac{1}{3}\Gamma\left(\frac{5}{3}\right) = \frac{1}{3}(0.89) \approx 0.297.$$

6. For $x > 0$

$$\Gamma(x+1) = \int_0^\infty t^x e^{-t} dt$$

$u = t^x$	$dv = e^{-t}\,dt$
$du = xt^{x-1}\,dt$	$v = -e^{-t}$

$$= -t^x e^{-t}\Big|_0^\infty - \int_0^\infty xt^{x-1}(-e^{-t})\,dt$$

$$= x\int_0^\infty t^{x-1}e^{-t}dt = x\Gamma(x).$$

Appendix II

3. (a) $\mathbf{AB} = \begin{pmatrix} -2-9 & 12-6 \\ 5+12 & -30+8 \end{pmatrix} = \begin{pmatrix} -11 & 6 \\ 17 & -22 \end{pmatrix}$

(b) $\mathbf{BA} = \begin{pmatrix} -2-30 & 3+24 \\ 6-10 & -9+8 \end{pmatrix} = \begin{pmatrix} -32 & 27 \\ -4 & -1 \end{pmatrix}$

(c) $\mathbf{A}^2 = \begin{pmatrix} 4+15 & -6-12 \\ -10-20 & 15+16 \end{pmatrix} = \begin{pmatrix} 19 & -18 \\ -30 & 31 \end{pmatrix}$

(d) $\mathbf{B}^2 = \begin{pmatrix} 1+18 & -6+12 \\ -3+6 & 18+4 \end{pmatrix} = \begin{pmatrix} 19 & 6 \\ 3 & 22 \end{pmatrix}$

6. (a) $\mathbf{AB} = (5 \quad -6 \quad 7)\begin{pmatrix} 3 \\ 4 \\ -1 \end{pmatrix} = (-16)$

(b) $\mathbf{BA} = \begin{pmatrix} 3 \\ 4 \\ -1 \end{pmatrix} (5 \quad -6 \quad 7) = \begin{pmatrix} 15 & -18 & 21 \\ 20 & -24 & 28 \\ -5 & 6 & -7 \end{pmatrix}$

(c) $(\mathbf{BA})\mathbf{C} = \begin{pmatrix} 15 & -18 & 21 \\ 20 & -24 & 28 \\ -5 & 6 & -7 \end{pmatrix} \begin{pmatrix} 1 & 2 & 4 \\ 0 & 1 & -1 \\ 3 & 2 & 1 \end{pmatrix} = \begin{pmatrix} 78 & 54 & 99 \\ 104 & 72 & 132 \\ -26 & -18 & -33 \end{pmatrix}$

(d) Since $\mathbf{AB}$ is 1×1 and $\mathbf{C}$ is 3×3 the product $(\mathbf{AB})\mathbf{C}$ is not defined.

9. (a) $(\mathbf{AB})^T = \begin{pmatrix} 7 & 10 \\ 38 & 75 \end{pmatrix}^T = \begin{pmatrix} 7 & 38 \\ 10 & 75 \end{pmatrix}$

(b) $\mathbf{B}^T\mathbf{A}^T = \begin{pmatrix} 5 & -2 \\ 10 & -5 \end{pmatrix} \begin{pmatrix} 3 & 8 \\ 4 & 1 \end{pmatrix} = \begin{pmatrix} 7 & 38 \\ 10 & 75 \end{pmatrix}$

12. $\begin{pmatrix} 6t \\ 3t^2 \\ -3t \end{pmatrix} + \begin{pmatrix} -t+1 \\ -t^2+t \\ 3t-3 \end{pmatrix} - \begin{pmatrix} 6t \\ 8 \\ -10t \end{pmatrix} = \begin{pmatrix} -t+1 \\ 2t^2+t-8 \\ 10t-3 \end{pmatrix}$

15. Since $\det \mathbf{A} = 0$, $\mathbf{A}$ is singular.

18. Since $\det \mathbf{A} = -6$, $\mathbf{A}$ is nonsingular.

$$\mathbf{A}^{-1} = -\frac{1}{6} \begin{pmatrix} 2 & -10 \\ -2 & 7 \end{pmatrix}$$

21. Since $\det \mathbf{A} = -9$, $\mathbf{A}$ is nonsingular. The cofactors are

$$\begin{aligned} A_{11} &= -2 & A_{12} &= -13 & A_{13} &= 8 \\ A_{21} &= -2 & A_{22} &= 5 & A_{23} &= -1 \\ A_{31} &= -1 & A_{32} &= 7 & A_{33} &= -5. \end{aligned}$$

Then

$$\mathbf{A}^{-1} = -\frac{1}{9} \begin{pmatrix} -2 & -13 & 8 \\ -2 & 5 & -1 \\ -1 & 7 & -5 \end{pmatrix}^T = -\frac{1}{9} \begin{pmatrix} -2 & -2 & -1 \\ -13 & 5 & 7 \\ 8 & -1 & -5 \end{pmatrix}.$$

24. Since $\det \mathbf{A}(t) = 2e^{2t} \neq 0$, $\mathbf{A}$ is nonsingular.

$$\mathbf{A}^{-1} = \frac{1}{2}e^{-2t} \begin{pmatrix} e^t \sin t & 2e^t \cos t \\ -e^t \cos t & 2e^t \sin t \end{pmatrix}$$

27. $\mathbf{X} = \begin{pmatrix} 2e^{2t} + 8e^{-3t} \\ -2e^{2t} + 4e^{-3t} \end{pmatrix}$ so that $\dfrac{d\mathbf{X}}{dt} = \begin{pmatrix} 4e^{2t} - 24e^{-3t} \\ -4e^{2t} - 12e^{-3t} \end{pmatrix}$.

30. **(a)** $\dfrac{d\mathbf{A}}{dt} = \begin{pmatrix} -2t/(t^2+1)^2 & 3 \\ 2t & 1 \end{pmatrix}$

(b) $\dfrac{d\mathbf{B}}{dt} = \begin{pmatrix} 6 & 0 \\ -1/t^2 & 4 \end{pmatrix}$

(c) $\displaystyle\int_0^1 \mathbf{A}(t)\,dt = \begin{pmatrix} \tan^{-1} t & \frac{3}{2}t^2 \\ \frac{1}{3}t^3 & \frac{1}{2}t^2 \end{pmatrix}\Bigg|_{t=0}^{t=1} = \begin{pmatrix} \frac{\pi}{4} & \frac{3}{2} \\ \frac{1}{3} & \frac{1}{2} \end{pmatrix}$

(d) $\displaystyle\int_1^2 \mathbf{B}(t)\,dt = \begin{pmatrix} 3t^2 & 2t \\ \ln t & 2t^2 \end{pmatrix}\Bigg|_{t=1}^{t=2} = \begin{pmatrix} 9 & 2 \\ \ln 2 & 6 \end{pmatrix}$

(e) $\mathbf{A}(t)\mathbf{B}(t) = \begin{pmatrix} 6t/(t^2+1) + 3 & 2/(t^2+1) + 12t^2 \\ 6t^3 + 1 & 2t^2 + 4t^2 \end{pmatrix}$

(f) $\dfrac{d}{dt}\mathbf{A}(t)\mathbf{B}(t) = \begin{pmatrix} (6 - 6t^2)/(t^2+1)^2 & -4t/(t^2+1)^2 + 24t \\ 18t^2 & 12t \end{pmatrix}$

(g) $\displaystyle\int_1^t \mathbf{A}(s)\mathbf{B}(s)\,ds = \begin{pmatrix} 6s/(s^2+1) + 3 & 2/(s^2+1) + 12s^2 \\ 6s^3 + 1 & 6s^2 \end{pmatrix}\Bigg|_{s=1}^{s=t}$

$$= \begin{pmatrix} 3t + 3\ln(t^2+1) - 3 - 3\ln 2 & 4t^3 + 2\tan^{-1} t - 4 - \pi/2 \\ (3/2)t^4 + t - (5/2) & 2t^3 - 2 \end{pmatrix}$$

33. $\left(\begin{array}{ccc|c} 1 & -1 & -5 & 7 \\ 5 & 4 & -16 & -10 \\ 0 & 1 & 1 & -5 \end{array}\right) \Longrightarrow \left(\begin{array}{ccc|c} 1 & -1 & -5 & 7 \\ 0 & 1 & 1 & -5 \\ 0 & 9 & 9 & -45 \end{array}\right) \Longrightarrow \left(\begin{array}{ccc|c} 1 & 0 & -4 & 2 \\ 0 & 1 & 1 & -5 \\ 0 & 0 & 0 & 0 \end{array}\right)$

Letting $z = t$ we find $y = -5 - t$, and $x = 2 + 4t$.

36. $\left(\begin{array}{ccc|c} 1 & 0 & 2 & 8 \\ 1 & 2 & -2 & 4 \\ 2 & 5 & -6 & 6 \end{array}\right) \Longrightarrow \left(\begin{array}{ccc|c} 1 & 0 & 2 & 8 \\ 0 & 2 & -4 & -4 \\ 0 & 5 & -10 & -10 \end{array}\right) \Longrightarrow \left(\begin{array}{ccc|c} 1 & 0 & 2 & 8 \\ 0 & 1 & -2 & -2 \\ 0 & 0 & 0 & 0 \end{array}\right)$

Letting $z = t$ we find $y = -2 + 2t$, and $x = 8 - 2t$.

39. $\left(\begin{array}{ccc|c}1 & 2 & 4 & 2\\ 2 & 4 & 3 & 1\\ 1 & 2 & -1 & 7\end{array}\right) \Longrightarrow \left(\begin{array}{ccc|c}1 & 2 & 4 & 2\\ 0 & 0 & -5 & -3\\ 0 & 0 & -5 & 5\end{array}\right) \Longrightarrow \left(\begin{array}{ccc|c}1 & 2 & 0 & -2/5\\ 0 & 0 & 1 & 3/5\\ 0 & 0 & 0 & 8\end{array}\right)$

There is no solution.

42. We solve

$$\det(\mathbf{A}-\lambda\mathbf{I}) = \begin{vmatrix}2-\lambda & 1\\ 2 & 1-\lambda\end{vmatrix} = \lambda(\lambda-3) = 0.$$

For $\lambda_1 = 0$ we have

$$\left(\begin{array}{cc|c}2 & 1 & 0\\ 2 & 1 & 0\end{array}\right) \Longrightarrow \left(\begin{array}{cc|c}1 & 1/2 & 0\\ 0 & 0 & 0\end{array}\right)$$

so that $k_1 = -\frac{1}{2}k_2$. If $k_2 = 2$ then

$$\mathbf{K}_1 = \begin{pmatrix}-1\\ 2\end{pmatrix}.$$

For $\lambda_2 = 3$ we have

$$\left(\begin{array}{cc|c}-1 & 1 & 0\\ 2 & -2 & 0\end{array}\right) \Longrightarrow \left(\begin{array}{cc|c}1 & -1 & 0\\ 0 & 0 & 0\end{array}\right)$$

so that $k_1 = k_2$. If $k_2 = 1$ then

$$\mathbf{K}_2 = \begin{pmatrix}1\\ 1\end{pmatrix}.$$

45. We solve

$$\det(\mathbf{A}-\lambda\mathbf{I}) = \begin{vmatrix}5-\lambda & -1 & 0\\ 0 & -5-\lambda & 9\\ 5 & -1 & -\lambda\end{vmatrix} = \begin{vmatrix}4-\lambda & -1 & 0\\ 4-\lambda & -5-\lambda & 9\\ 4-\lambda & -1 & -\lambda\end{vmatrix} = \lambda(4-\lambda)(\lambda+4) = 0.$$

If $\lambda_1 = 0$ then

$$\left(\begin{array}{ccc|c}5 & -1 & 0 & 0\\ 0 & -5 & 9 & 0\\ 5 & -1 & 0 & 0\end{array}\right) \Longrightarrow \left(\begin{array}{ccc|c}1 & 0 & -9/25 & 0\\ 0 & 1 & -9/5 & 0\\ 0 & 0 & 0 & 0\end{array}\right)$$

so that $k_1 = \frac{9}{25}k_3$ and $k_2 = \frac{9}{5}k_3$. If $k_3 = 25$ then

$$\mathbf{K}_1 = \begin{pmatrix}9\\ 45\\ 25\end{pmatrix}.$$

If $\lambda_2 = 4$ then

$$\left(\begin{array}{ccc|c} 1 & -1 & 0 & 0 \\ 0 & -9 & 9 & 0 \\ 5 & -1 & -4 & 0 \end{array}\right) \Longrightarrow \left(\begin{array}{ccc|c} 1 & 0 & -1 & 0 \\ 0 & 1 & -1 & 0 \\ 0 & 0 & 0 & 0 \end{array}\right)$$

so that $k_1 = k_3$ and $k_2 = k_3$. If $k_3 = 1$ then

$$\mathbf{K}_2 = \begin{pmatrix} 1 \\ 1 \\ 1 \end{pmatrix}.$$

If $\lambda_3 = -4$ then

$$\left(\begin{array}{ccc|c} 9 & -1 & 0 & 0 \\ 0 & -1 & 9 & 0 \\ 5 & -1 & 4 & 0 \end{array}\right) \Longrightarrow \left(\begin{array}{ccc|c} 1 & 0 & -1 & 0 \\ 0 & 1 & -9 & 0 \\ 0 & 0 & 0 & 0 \end{array}\right)$$

so that $k_1 = k_3$ and $k_2 = 9k_3$. If $k_3 = 1$ then

$$\mathbf{K}_3 = \begin{pmatrix} 1 \\ 9 \\ 1 \end{pmatrix}.$$

48. We solve

$$\det(\mathbf{A} - \lambda\mathbf{I}) = \begin{vmatrix} 1-\lambda & 6 & 0 \\ 0 & 2-\lambda & 1 \\ 0 & 1 & 2-\lambda \end{vmatrix} = \begin{vmatrix} 1-\lambda & 6 & 0 \\ 0 & 3-\lambda & 3-\lambda \\ 0 & 1 & 2-\lambda \end{vmatrix} = (3-\lambda)(1-\lambda)^2 = 0.$$

For $\lambda = 3$ we have

$$\left(\begin{array}{ccc|c} -2 & 6 & 0 & 0 \\ 0 & 0 & 0 & 0 \\ 0 & 1 & -1 & 0 \end{array}\right) \Longrightarrow \left(\begin{array}{ccc|c} 1 & 0 & -3 & 0 \\ 0 & 1 & -1 & 0 \\ 0 & 0 & 0 & 0 \end{array}\right)$$

so that $k_1 = 3k_3$ and $k_2 = k_3$. If $k_3 = 1$ then

$$\mathbf{K}_1 = \begin{pmatrix} 3 \\ 1 \\ 1 \end{pmatrix}.$$

For $\lambda_2 = \lambda_3 = 1$ we have

$$\left(\begin{array}{ccc|c} 0 & 6 & 0 & 0 \\ 0 & 1 & 1 & 0 \\ 0 & 1 & 1 & 0 \end{array}\right) \Longrightarrow \left(\begin{array}{ccc|c} 0 & 1 & 0 & 0 \\ 0 & 0 & 1 & 0 \\ 0 & 0 & 0 & 0 \end{array}\right)$$

so that $k_2 = 0$ and $k_3 = 0$. If $k_1 = 1$ then

$$\mathbf{K}_2 = \begin{pmatrix} 1 \\ 0 \\ 0 \end{pmatrix}.$$

51. Let

$$\mathbf{A} = \begin{pmatrix} a_{11} & a_{12} \\ a_{21} & a_{22} \end{pmatrix}.$$

Then

$$\frac{d}{dt}[\mathbf{A}(t)\mathbf{X}(t)] = \frac{d}{dt}\begin{pmatrix} a_1 & a_2 \\ a_3 & a_4 \end{pmatrix}\begin{pmatrix} x_1 \\ x_2 \end{pmatrix} = \frac{d}{dt}\begin{pmatrix} a_1x_1 + a_2x_2 \\ a_3x_1 + a_4x_2 \end{pmatrix} = \begin{pmatrix} a_1x_1' + a_1'x_1 + a_2x_2' + a_2'x_2 \\ a_3x_1' + a_3'x_1 + a_4x_2' + a_4'x_2 \end{pmatrix}$$

$$= \begin{pmatrix} a_1 & a_2 \\ a_3 & a_4 \end{pmatrix}\begin{pmatrix} x_1' \\ x_2' \end{pmatrix} + \begin{pmatrix} a_1' & a_2' \\ a_3' & a_4' \end{pmatrix}\begin{pmatrix} x_1 \\ x_2 \end{pmatrix} = \mathbf{A}(t)\mathbf{X}'(t) + \mathbf{A}'(t)\mathbf{X}(t).$$

54. Since

$$(\mathbf{AB})(\mathbf{B}^{-1}\mathbf{A}^{-1}) = \mathbf{A}(\mathbf{BB}^{-1})\mathbf{A}^{-1} = \mathbf{AIA}^{-1} = \mathbf{AA}^{-1} = \mathbf{I}$$

and

$$(\mathbf{B}^{-1}\mathbf{A}^{-1})(\mathbf{AB}) = \mathbf{B}^{-1}(\mathbf{A}^{-1}\mathbf{A})\mathbf{B} = \mathbf{B}^{-1}\mathbf{IB} = \mathbf{B}^{-1}\mathbf{B} = \mathbf{I}$$

we have

$$(\mathbf{AB})^{-1} = \mathbf{B}^{-1}\mathbf{A}^{-1}.$$